Windows on the Wild®

Biodiversity Basics
Student Book

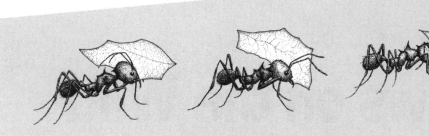

Welcome to Biodiversity Basics!

This Student Book contains all the student pages that link to the activities in *Biodiversity Basics—An Educator's Guide to Exploring the Web of Life*. We created a separate book to make it easier for you to copy the articles, stories, puzzles, worksheets, and cartoons that you'll need to conduct the activities in the module.

As you'll see, all the pages in this volume are designed to be reproduced. Sometimes you may want to make a copy for every student; other times, you may only need one for every small group. And if you're working with a small group of students or children in a home learning situation, you might want to have a copy of the Student Book for each person.

These activity pages are arranged in the same order as *Biodiversity Basics* and grouped into the same four chapters. Each ready-to-copy page has its own title, but also includes the title and number of the activity it belongs with. The general directions for how to use each page are included in the Educator's Guide, although when appropriate, some of the pages in this book also have student directions.

We hope these student pages make your job a little easier by increasing your teaching options and saving you time. And we'd love to hear how you adapt these pages (and the activities) so that we can share your ideas with other educators.

Good luck, and check out our Web site for more ideas.

www.worldwildlife.org

TABLE OF CONTENTS

CHAPTER 2 Why Is Biodiversity Important? 61

13 Super Sleuths

14 The F-Files

15 Biodiversity Performs!

16 The Nature of Poetry

18 The Culture/Nature Connection

19 Secret Services

20 Diversity on Your Table

What's the Status of Biodiversity? 111

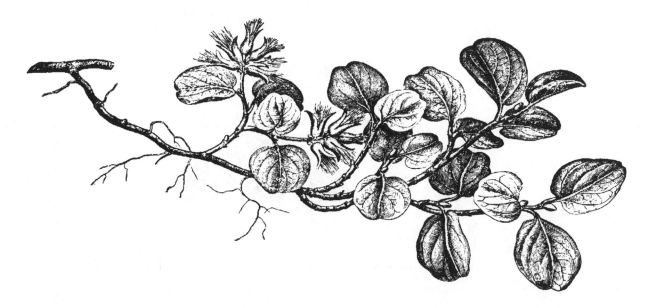

What Is Biodiver

The activity pages in this section introduce the concept of biodiversity and highlight the important roles biodiversity plays in our lives. For the corresponding activities, see pages 72-179 in the Educator's Guide.

sity?

Pete Oxford/ENP Images

"*Earth is home to tens of millions of living species, of which we are but one—a legacy of 3.5 million years of evolution. Biodiversity is the spectacular variety of life on Earth and the essential interdependence among all living things.*"

—Michael Novacek, biologist

Here's your chance to find out what you know about the world's diverse plants, animals, and natural places. For each question, circle <u>all</u> the correct answers.

1. Which of the following could the fastest human outrun in a 100-yard race?

    a. cheetah

    b. wart hog

    c. three-toed sloth

    d. domestic cat

    e. wild turkey

2. Which of the following actually exist?

    a. ants that "herd" aphids for food

    b. slime molds that creep across the ground

    c. flowers that "trick" insects into mating with them

    d. none of the above

3. What does a large adult bluefin tuna have in common with a Porsche 911?

    a. ability to go from 0 to 60 MPH in 4.7 seconds

    b. big fins

    c. a sticker price of about $60,000

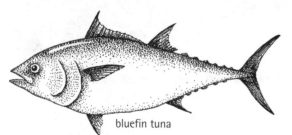

bluefin tuna

4. Which of the following best describes the word biodiversity?

    a. endangered species

    b. different kinds of planets in the solar system

    c. the variety of all life on Earth

    d. a bunch of biographies about famous biologists

5. U. S. Fish and Wildlife Service agents and U. S. Customs officials in Los Angeles, California, once found which of the following strapped to a man's arms and legs?

    a. 18 California king snakes

    b. 45 pounds of elephant ivory

    c. 16 sea turtle eggs

    d. 16 vampire bats

6. Scientists studying bug zappers have learned some interesting facts. Which of the following are among them?

    a. Insects are attracted to bug zappers because of the zappers' smoky smell.

    b. Bug zappers are great for ridding summer nights of mosquitoes.

    c. Bug zappers could be bad news for certain birds, fish, bats, and flowers.

    d. There are more than 4 million bug zappers being used in the United States.

7. Blackpoll warblers are tiny birds that migrate between North America and South America each year. Which of the following statements about them are true?

    a. They use the stars for navigation.

    b. They make frequent pit stops at fast-food restaurants.

    c. They don't really need to migrate.

    d. If they burned gasoline instead of body fat for fuel, they'd get 720,000 miles to the gallon.

WHAT'S YOUR BIODIVERSITY IQ? (Cont'd.)

8. Which of the following can be considered an enemy of coral reefs?

    a. starfish

    b. jewelry

    c. sunken treasure

    d. divers

9. What's the most serious threat to biodiversity?

    a. sharks

    b. habitat loss

    c. tourists

    d. pollution

10. The items on the left have been (or are being) developed into important medicines for humans. Match each item with the medicine it inspired by writing the letters in the appropriate blanks.

___ bread mold	**a.** anti-inflammatory
___ white willow tree	**b.** antibiotic
___ vampire bat saliva	**c.** pain reliever
___ wild yams	**d.** medicine to unclog arteries
___ Caribbean coral	**e.** first-aid ointment

11. Which of the following are true statements about camels?

    a. They store water in their humps.

    b. During cooler weather, they can go up to two months without drinking.

    c. They played a key role in opening up trade across the deserts of Asia and Africa.

    d. They provide people with milk, meat, cooking fuel (in the form of dried dung), wool, and leather.

12. Without fungi, which of the following would you not be able to do?

    a. eat pizza topped with pepperoni and mushrooms

    b. bake bread

    c. live in a world free of dead things lying all over the place

    d. put blue cheese dressing on your salad

13. Which of the following statements are true?

    a. Potatoes originated in Ireland.

    b. The United States grows most of its baking potatoes in Washington.

    c. More than 5,000 different kinds of potatoes have been identified in South America's Andes Mountains.

    d. The French fry, invented by Madame Bonaparte during the French Revolution, became one of Napoleon's favorite snacks.

dromedary and Bactrian camel

14. Which of the following are real species of animals?

   a. vampire moths

   b. Tasmanian devils

   c. Komodo dragons

   d. werewolves

15. If you decided to throw a party to celebrate the diversity of life on Earth and wanted to send an invitation to each species, how many invitations would you need?

   a. 150

   b. about 3,000

   c. 652,983

   d. more than 1.5 million

16. Which of the following statements about ostriches are true?

   a. They hide their heads in the sand.

   b. Their meat is so tasty that some people are choosing it over beef for burgers.

   c. They fly when no one is looking.

   d. They can outrun a lion or a hyena.

17. If the number of species on Earth was represented by physical size, which of the following would most accurately illustrate the proportion of insects to mammals?

   a.

   b.

   c.

18. Biodiversity includes:

   a. the color of your eyes

   b. the creatures in your neighborhood soil

   c. Antarctica

   d. your classmates

19. If there were a prize for "the strongest creature for its size," which of the following would win?

   a. gorillas

   b. chickens

   c. ants

   d. turtles

20. Which of the following would people have to do without if there were no bees?

   a. almonds

   b. honey

   c. cucumbers

   d. apples

   e. celery

21. Which of the following is an example of an ecosystem service?

   a. a ladybug that protects your garden by eating aphid pests

   b. a company that rakes people's yards

   c. a wetland that filters dirty water

   d. an ocean that controls the Earth's climate

WHAT'S YOUR BIODIVERSITY IQ? (Cont'd.)

22. Some of the world's most fascinating creatures live in really unusual places. Which of the following is sometimes a home for another living thing?

    a. a lobster's mouth
    b. a termite's gut
    c. a rhino's gut
    d. a human's forehead

one-horned rhino

23. If you had a job that put you in charge of saving all species on the edge of extinction in the United States, about how many endangered species would you need to save (based on what we know today)?

    a. 12
    b. 250
    c. 517
    d. 1,082

24. A small population of beluga whales lives in Canada's St. Lawrence River. Which of the following explains why they are threatened?

    a. In winter the river freezes, making it nearly impossible for the whales to surface for air.
    b. Beluga steaks are considered a delicacy in many parts of the world.
    c. The areas of the river where the whales feed are heavily polluted, and as a result the whales are contaminated with toxins.
    d. The river's water is really too shallow for such large mammals.

25. Which of the following environments on our planet are too harsh to support life?

    a. boiling sulfur springs, where temperatures are commonly 212° Fahrenheit (100° Celcius)
    b. deep-sea hydrothermal vents called "smokers," where the temperature can reach 662° Fahrenheit (350° Celsius)
    c. the frigid ice of the Arctic and Antarctic
    d. all of the above
    e. none of the above

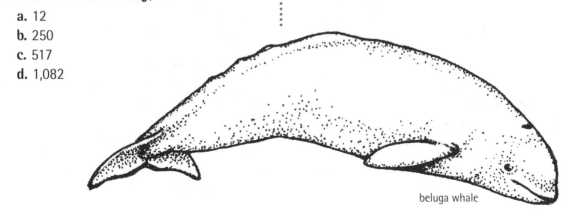

beluga whale

"**M**an, this is getting weird," said Jehan under her breath. She stared at the small piece of paper that had fluttered out onto the floor when she opened her locker. Six mysterious words, scrawled in bright green ink, stared back at her.

You can't get away from it! blasted the green letters. Jehan felt her skin crawl. It was the third note of its kind she'd gotten in two days, and they were starting to creep her out.

She'd found the first one in her geometry book yesterday after lunch. **Your very life depends on it!** the note had announced. At the time, Jehan hadn't paid much attention to it. *Whoever owned this book last year didn't get all their papers out*, she thought, pulling out the note to add it to the recycling box. But for some reason—she wasn't sure why—she crammed it into the back of her notebook instead.

Later that day Jehan had found another message, this time in her jacket pocket. She frowned slightly as she read its menacing words: **Ignoring it could be a fatal mistake!** Just what was *that* supposed to mean? Was it some kind of threat? Jehan impatiently stuffed the second note into the back of her notebook next to the first one.

Now, looking over her shoulder, Jehan bent down to pick up the third note. *OK, who's trying to freak me out?* she wondered. Just then she saw her friend Megan coming toward her. "Aha," muttered Jehan, folding her arms against her chest. Weird notes were just the kind of thing Megan's warped sense of humor would dream up.

"Hey, J, check this out," said Megan, thrusting a piece of paper at Jehan. "I just found it in my locker. It's the third note I've gotten since yesterday."

Your pizza depends on it! said the note cryptically.

Jehan's eyes narrowed. "Oh, right," she said. "I'm supposed to think you have nothing to do with these notes just because you're pretending to get them, too."

"Too?" said Megan, looking surprised. "You mean I'm not the only one? Well, that's a relief."

Jehan glanced suspiciously at her friend, but Megan didn't have that I'm-so-innocent look she always wore when she was playing one of her dumb jokes. In fact, she really *did* look innocent.

Jehan sighed and showed Megan the three notes she'd gotten. Megan whistled under her breath. "'Fatal mistake!' 'Can't get away from it!' 'Your life depends on it!'" she read dramatically. "At least only my *pizza* depends on it, whatever 'it' is. So who's got it in for you, J?"

Good question, thought Jehan. *For that matter, why did Megan get a note about pizza, of all things? What did that have to do with anything?* She shook her head. Nothing was making sense.

"You should've seen the one I got yesterday," Megan said as they headed toward English class. "It said something like, **'Without it, nine-nut crackling crunch bars wouldn't even exist!'** Apparently my incredible crunch bars are so famous they're even mentioned in anonymous notes," she said, sighing. Megan was a great cook and she knew it.

"Wait a minute," said Jehan, stopping in her tracks. "Pizza and crunch bars . . . what did your third note say?" She had a feeling she was on to something.

"It was totally bizarre, something like **'Leetown High cafeteria food would be even worse without it!'**" Megan made a face. "Hard to imagine," she said.

"I knew it. All of your notes have something to do with food. And all of mine sound like something out of a Dr. Detecto story," said Jehan. She loved mysteries and suspense thrillers, especially the comic book series called *Dr. Detecto, Private Eye*. She couldn't get enough of Dr. Detecto's dumb adventures. The investigator bumbled his way through all kinds of mysteries, but he always managed to solve them.

Megan wasn't following Jehan's train of thought. "So what? What's pizza got to do with Dr. Detecto?" she asked.

"It's not what pizza has to do with Dr. Detecto. It's the fact that whoever wrote the notes knows something about us. "

Jehan smiled. She was beginning to feel a little like a detective herself. *OK, how would Dr. Detecto solve this mystery?* she wondered secretly. First, she figured, he'd collect more clues.

At lunch, Jehan got a chance to do just that. As she and Megan picked at something that was supposed to be lasagna, their friends Noah and Jamal joined them. A small piece of paper fluttered out of Noah's jacket pocket as he draped the jacket over a chair. "Hey, what's this?" he said, picking the paper up. It was another note. Noah said it was the third one he'd gotten since yesterday. Jamal said he'd also gotten three notes.

Excellent, thought Jehan, pulling a pencil and small notebook out of her backpack. "So what did your notes say?" she asked the guys.

Just as Jehan had suspected, each of their notes focused on a topic that they were especially interested in. Noah's all had something to do with sports. The one he had just found said, **Without it, there would be no Detroit Tigers.** Another claimed, **The future of the Baltimore Orioles could depend on it!** And the third note said, **If it didn't exist, you wouldn't be a member of the Leetown Leopards junior varsity basketball team.**

Jamal, who had decided when he was seven years old that he was going to be a doctor, got notes having to do with medicine. **It could provide a cure for cancer!** announced one. **It's the source of many important prescription drugs,** read another. **It's the world's largest pharmacy!** stated the third.

Jehan copied down each note. "Hmmm. If the four of us are getting notes, I bet other kids are, too," she said. "The next step is to find out who."

"How are you going to do that, Dr. Detecto?" asked Jamal, tapping his pen on the table. He had a funny look on his face, as if he was extremely amused by the whole thing.

"I'll think of something," answered Jehan. In fact, she was already thinking of something. She was thinking of her friend Nikki, who was in charge of making the daily announcement over the intercom.

That afternoon, all of Leetown High heard the following announcement: **Have you been getting strange, anonymous notes over the past few days? If so, come help solve the mystery by meeting in the study area tomorrow, right after seventh period.**

When Jehan got to the study area the next day, Jamal was already there. Soon Noah and Megan showed up. And over the next few minutes, at least ten other kids came in. Jehan called the meeting to order and asked the kids to read their notes. There were some strange ones, all right:

- **Not hundreds. Not even thousands. We're talking millions!**
- **A vampire could save your life!**
- **Aliens are among us!**
- **Don't look now, but there may be something fishy about your eye shadow.**
- **Bugs rule!**

"OK," said Jehan, "looks like we've got everything from bugs to pizza to eye shadow here. That covers a lot of territory. I have no idea what it all means, but I have a feeling there's someone here who does. Jamal, how about giving us a clue?"

Jamal looked pleasantly surprised. He smiled at Jehan, stood up, and said, "Welcome to the third meeting of BioForce—the Leetown Biodiversity Club. Thanks for getting everyone together, Jehan. You saved me the trouble!"

One purpose of the club, explained Jamal, was to explore the incredible variety of life on Earth—called biodiversity—and the different ways it affects people's lives. Another was to take action to help protect biodiversity, both locally and around the world.

"Biodiversity has something for everyone," said Jamal. "Whether you're interested in sports or art or science or animals or food or social issues or—," he grinned at Jehan, "—whether you just like a good mystery."

"There's plenty of *that* in these notes," said Megan. "I mean, what could pizza possibly have to do with biodiversity?"

"It's all right here," said Jamal, passing out copies of a booklet called *The BioForce Fact Book*. "If you read through this you'll find out what all your notes mean."

Jamal explained that the booklet was the first project of the founding members of the club. They had designed it on the computer and planned to sell copies to raise money for a local charity. Other projects in the works included a food festival, featuring ingredients from

biodiversity "hot spots" around the world ("Count me in," said Megan); a display for the local community library on the link between toxins and health issues (Jamal was in charge of that project); and an undercover investigation of local pet stores to see whether tropical parrots and aquarium fish had been illegally taken from the wild. (Jehan volunteered to lead the investigation.)

Later, after the next BioForce meeting had been set and the current meeting adjourned, Jamal and Jehan walked out to catch the late bus together.

"So, how did you know I was the one behind all those notes?" asked Jamal.

"Oh, please. It was a no-brainer," laughed Jehan.

Jamal raised his eyebrows. "No way," he said. "Not a single person saw me planting those notes! I was amazingly sneaky!"

"Sneaky, yes, but you're no match for a natural detective like me," smirked Jehan. "First, you practically waved your green pen under my nose at lunch yesterday. I figured it was the same one the notes were written with, since they were in bright green ink. Second, who else would have known enough about all that medical stuff to come up with the three notes you pretended to get? And third, you were the first one at the meeting today. That was the real clincher. You're never on time for anything!"

"OK, so maybe it was an open-and-shut case," grinned Jamal. "But you're definitely tough to fool. Nice work, Dr. Detecto."

MYSTERY NOTES

Match these notes from "Mystery at Leetown High" with one or more facts from the "BioForce Fact Book" by writing the number of the fact in the blank next to the note or notes it "fits." (Each fact can fit more than one note. And each note can have more than one fact.) Be able to explain how the facts fit the notes.

Jehan's Notes

_____ You can't get away from it!

_____ Your very life depends on it!

_____ Ignoring it could be a fatal mistake!

Megan's Notes

_____ Your pizza depends on it!

_____ Without it, nine-nut crackling crunch bars wouldn't even exist!

_____ Leetown High cafeteria food would be even worse without it!

Noah's Notes

_____ Without it, there would be no Detroit Tigers.

_____ The future of the Baltimore Orioles could depend on it!

_____ If it didn't exist, you wouldn't be a member of the Leetown Leopards junior varsity basketball team.

Jamal's Notes

_____ It could provide a cure for cancer!

_____ It's the source of many important prescription drugs.

_____ It's the world's largest pharmacy!

Miscellaneous Notes

_____ Not hundreds. Not even thousands. We're talking millions!

_____ A vampire could save your life!

_____ Aliens are among us!

_____ Don't look now, but there may be something fishy about your eye shadow.

_____ Bugs rule!

1. When it comes to knowing what kinds of life forms are out there, scientists may have only scratched the surface. So far they've identified about 1.7 million species—but there may be as many as 100 million! All those species, along with the different ecosystems they live in and the billions of genes they contain, make up the variety of life called biodiversity.

2. Biodiversity is everywhere! Planet Earth is literally crawling with life, from tiny bacteria in the soil under your feet to gigantic whales as long as a city block.

3. Lions and tigers and bears—and blue jays and cardinals and eagles! These animals are just a few of the ones that sports teams are named after. There are dozens of others. Animals inspire us not only in sports but also in many other areas of our lives.

4. Biodiversity is much more than a source of names for sports teams. For example, scientists are learning that the millions of species on Earth, along with the ecosystems they live in, have a role to play in supporting one another. Because of this, many scientists and others feel that human well-being depends on keeping Earth's biodiversity healthy.

5. The natural world is one big medicine chest. One-fourth of all prescription drugs used today were originally derived from plants. And scientists think that many more are waiting to be discovered among the rich biodiversity of life on Earth.

6. One of the biggest threats to the biodiversity of many areas is "alien" species—species that humans, either accidentally or purposefully, have introduced into places where they didn't occur before. (Norway rats, the kudzu vine, and starlings [a type of bird] are a few of the thousands of examples.) These species often thrive in their new homes because they may not have many predators in the areas they're introduced into. As a result, alien species can "take over" their new habitats and push out species that were already there. (Scientists also call alien species *introduced* or *exotic* species.)

7. You can thank biodiversity for your favorite edible fungus, whether it's the "blue" in your blue cheese salad dressing or the mushrooms that mingle with the olives and pepperoni on top of that popular Italian-American cheese and tomato-sauce pie.

8. There's a lot of beauty out there in biodiversity land—colorful coral reef fish, majestic eagles, cuddly looking pandas . . . . So why, you might ask, did nature have to ruin things by coming up with bugs, slugs, and other less-than-beautiful creatures? Well, you should be glad it did. Even the ugliest, weirdest, and scariest species have a role to play in nature. And many are proving valuable to people in ways no one could have guessed. Take vampire bats. Scientists have found a way to use their saliva to dissolve dangerous blood clots in humans. Not even the cuddliest, cutest panda can do that!

9. We humans are totally outnumbered. So are the rest of our fellow mammals. For that matter, so are reptiles, amphibians, and fish. That's because the total number of vertebrate species (those with backbones) doesn't come close to the number of insect species. There are more than 950,000 known species of insects, and scientists think the total number may be tens of millions. Aren't you glad insects are smaller than we are?

10. People are helping to preserve areas rich in biodiversity by growing and harvesting crops in ways that don't harm the habitats they're grown in. For example, in some tropical areas, farmers grow Brazil nuts, cashews, and other crops for export to North America, Europe, and elsewhere. By providing a product that people want and by using careful growing and harvesting methods, they ensure the protection of the areas' biodiversity.

11. Rain forests and other natural areas could have medical marvels within their midst—in the form of a cure for cancer, AIDS, or other life-threatening diseases. Scientists have stepped up their efforts to learn about the medicinal plants in these areas. In some places, they're learning a lot from shamans and other native healers.

12. The loss of tropical rain forests and other natural areas results in a loss in biodiversity—and not necessarily just in those areas. For example, recent studies suggest that populations of certain birds that migrate back and forth from North America to the tropics are decreasing because of the loss of rain forests and other tropical habitats. Some of our most colorful songbirds, including Northern orioles (formerly known as Baltimore orioles), are among the casualties. Habitat loss is the main reason we're losing biodiversity.

13. OK, so cafeteria food isn't exactly a gourmet treat. But at least you can choose what you're going to eat from a variety of different foods. If the thought of forcing down those soggy-looking Brussels sprouts makes you gag, you can choose the macaroni and cheese. Now consider for a moment what your school lunch might be like if the Earth didn't have much biodiversity. Yikes! Those overcooked Brussels sprouts might be the only thing on the school lunch menu! Oh, wait—there's always the salad bar. Of course, all it has on it is iceberg lettuce. Oh, well—bon appétit!

14. People list many reasons for wanting to protect biodiversity. Some feel that the quality of human life is richer simply because we're surrounded by so many other life forms. They say that this is reason enough to protect biodiversity. Others believe that we depend on biodiversity for our very survival. They feel that, without the complex interactions of all those species and the ecosystems they're a part of, everything from our atmosphere to the planet's oceans—not to mention humanity itself—could suffer the consequences.

15. Did you know that the red color in some face powders, lipsticks, blushes, and eye makeup comes from the bodies of crushed insects? And the luminous, translucent look of some nail polish comes from ground-up fish scales? Biodiversity works in mysterious ways.

Mystery 1

A 55-year-old woman is rushed into the emergency room. Two days later, she checks out of the hospital wearing an "I LOVE VAMPIRE BATS" button on her sweater. WHY?

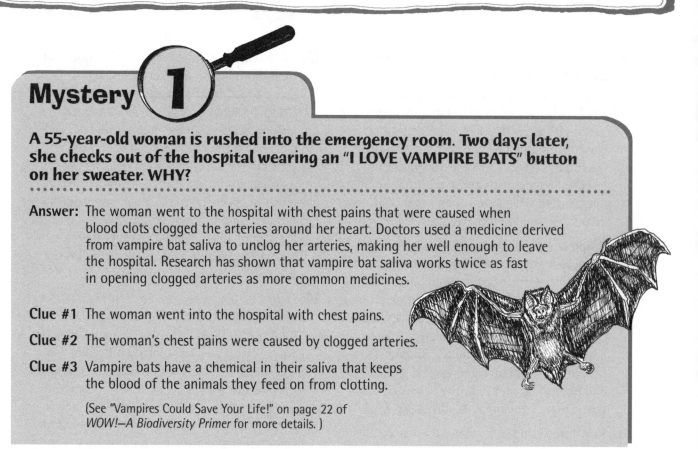

Answer: The woman went to the hospital with chest pains that were caused when blood clots clogged the arteries around her heart. Doctors used a medicine derived from vampire bat saliva to unclog her arteries, making her well enough to leave the hospital. Research has shown that vampire bat saliva works twice as fast in opening clogged arteries as more common medicines.

Clue #1 The woman went into the hospital with chest pains.

Clue #2 The woman's chest pains were caused by clogged arteries.

Clue #3 Vampire bats have a chemical in their saliva that keeps the blood of the animals they feed on from clotting.

(See "Vampires Could Save Your Life!" on page 22 of *WOW!—A Biodiversity Primer* for more details.)

Mystery 2

A woman modeling a new line of bright, sequined evening wear walks down a runway and wears matching, shimmering nail polish on her fingernails. "Your new look is terrific," a critic tells the clothing designer. "Thanks," replies the designer. "I call it my Fish Line." WHY?

Answer: The designer used ground-up fish scales to give the model's sequins and nail polish their iridescent, sparkly look. (While many commercial nail polishes are made from crushed fish scales, most sequins these days are plastic.)

Clue #1 The woman's sequins and nail polish have something in common.

Clue #2 Both the sequins and nail polish are very sparkly.

Clue #3 Fish are covered with thin, "sparkly" scales.

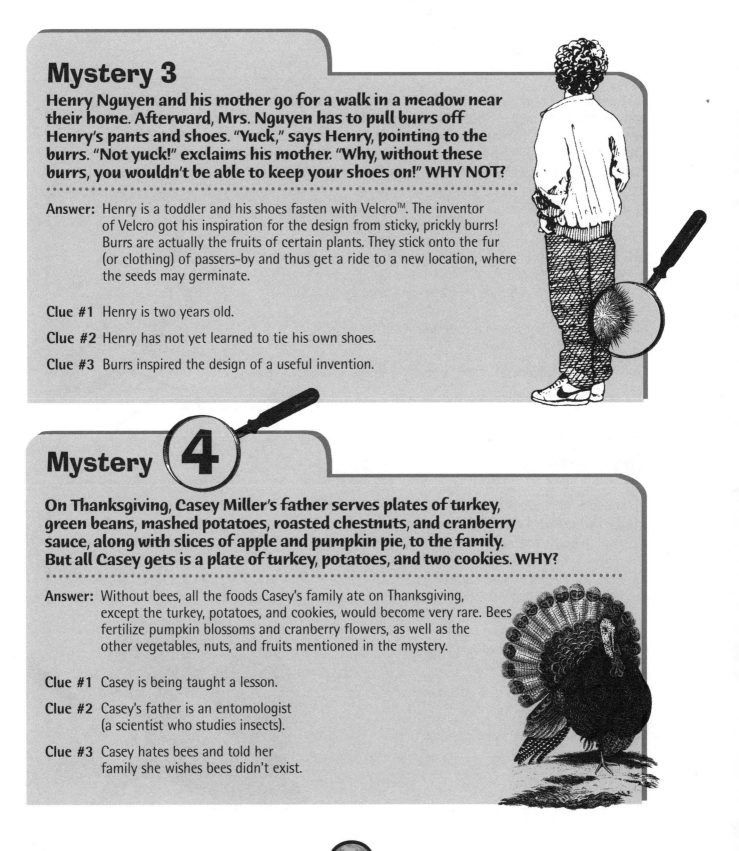

Mystery 3

Henry Nguyen and his mother go for a walk in a meadow near their home. Afterward, Mrs. Nguyen has to pull burrs off Henry's pants and shoes. "Yuck," says Henry, pointing to the burrs. "Not yuck!" exclaims his mother. "Why, without these burrs, you wouldn't be able to keep your shoes on!" WHY NOT?

Answer: Henry is a toddler and his shoes fasten with Velcro™. The inventor of Velcro got his inspiration for the design from sticky, prickly burrs! Burrs are actually the fruits of certain plants. They stick onto the fur (or clothing) of passers-by and thus get a ride to a new location, where the seeds may germinate.

Clue #1 Henry is two years old.

Clue #2 Henry has not yet learned to tie his own shoes.

Clue #3 Burrs inspired the design of a useful invention.

Mystery **4**

On Thanksgiving, Casey Miller's father serves plates of turkey, green beans, mashed potatoes, roasted chestnuts, and cranberry sauce, along with slices of apple and pumpkin pie, to the family. But all Casey gets is a plate of turkey, potatoes, and two cookies. WHY?

Answer: Without bees, all the foods Casey's family ate on Thanksgiving, except the turkey, potatoes, and cookies, would become very rare. Bees fertilize pumpkin blossoms and cranberry flowers, as well as the other vegetables, nuts, and fruits mentioned in the mystery.

Clue #1 Casey is being taught a lesson.

Clue #2 Casey's father is an entomologist (a scientist who studies insects).

Clue #3 Casey hates bees and told her family she wishes bees didn't exist.

World Wildlife Fund

What Is Biodiversity?

Mystery **5**

Roberto Fernandez goes with his sister Isabel to their local farmer's market to buy organic vegetables. Arriving at the market, they realize they don't yet know which farmers are the organic ones. As they look around at the different stands, Roberto sees a sign for a farm called "Mantis Meadows." "Oh, there!" he exclaims. "I bet that farmer is organic." WHY?

Answer: The praying mantis is an insect that eats other insects, including farm pests. For that reason, praying mantids are often encouraged by organic farmers, who do not use chemical pesticides. When Roberto saw a farm called "Mantis Meadows," he figured that was the place using nonchemical means to help control pests.

Clue #1 Insects that feed on crops are one of the biggest problems for farmers.

Clue #2 Organic farmers do not use chemical pesticides to kill insects.

Clue #3 Praying mantids eat lots of insects.

Mystery 6

Theresa Lee is watching TV with her sister Alice when a commercial for cold medicine comes on. In the commercial, four koalas sneeze and cough before taking the advertised remedy. "That's appropriate!" Theresa says. WHY?

Answer: The commercial is advertising a cough medicine made from eucalyptus leaves. Eucalyptus is also a favorite food of koalas.

Clue #1 The advertised medicine is especially designed to soothe coughs.

Clue #2 Many ingredients in cough medicine are extracted from plants.

Clue #3 This cough medicine contains eucalyptus leaves.

Mystery **7**

Ravi's dad is reading the paper one day. One of the headlines reads, "Hundreds Sick at Memorial Hospital—Bacteria Contamination Suspected." Ravi's dad turns to Ravi and says, "Sounds like the hospital might need some horseshoe crabs." WHY?

Answer: Horseshoe crab blood contains special cells that can be used to test for gram-negative bacteria, a certain type of bacteria that can contaminate disposable syringes, medical equipment, vaccines and other drugs, and intravenous fluids. Gram-negative bacteria are found all over the place, including our own intestines. But if they get into people's blood they can be deadly. That's because gram-negative bacteria produce lethal toxins that cause a burning fever. The discovery about horseshoe crab blood, made in 1964, led to the development of a test for gram-negative bacteria that was much faster and more accurate than the test people had been using. Not only did this test make it easier to screen vaccines, intravenous fluids, and medical equipment for gram-negative bacteria, but it also made it possible for doctors to diagnose whether patients were infected by the bacteria. And development of this test made horseshoe crabs one of the most valuable marine animals around. One liter of processed horseshoe crab blood is worth up to $8,000!

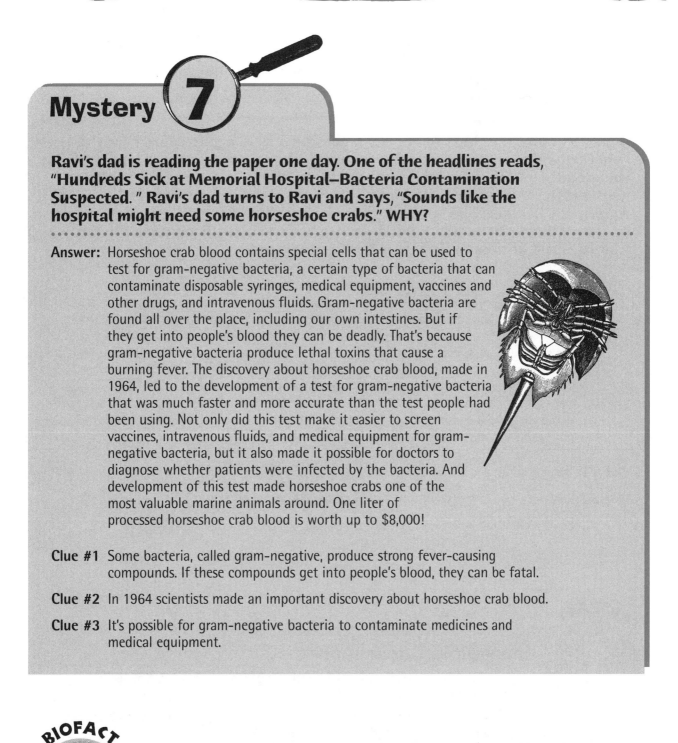

Clue #1 Some bacteria, called gram-negative, produce strong fever-causing compounds. If these compounds get into people's blood, they can be fatal.

Clue #2 In 1964 scientists made an important discovery about horseshoe crab blood.

Clue #3 It's possible for gram-negative bacteria to contaminate medicines and medical equipment.

BIOFACT

A Tree Full of Life—Dr. Terry Erwin, an entomologist studying insects in the rain forest, discovered that a single tree in some of the rain forests of South America can be home to more than 1,000 species of insects.

Mystery 8

A group of bird watchers in Michigan goes on a field trip to search for the rare Kirtland's warbler. When they arrive at their destination the leader of the group says, "There must be some mistake. I was told we'd find Kirtland's warblers here—but there's no way." "Why?" ask the disappointed bird watchers. "Because," he answers, "it looks like there hasn't been a fire in this area for at least 20 years." What does fire have to do with whether the bird watchers will be able to find their feathered friend?

Answer: Kirtland's warbler, an endangered species, lives only in jack pine forests in central Michigan. Periodic wild fires play an important role in maintaining these forests. When a fire sweeps through a stand of mature jack pines, the heat of the fire causes the trees' pine cones to open and release their seeds into the nutrient-rich ash left behind. When the saplings that sprout in this soil get to be between five and twelve feet tall, Kirtland's warblers nest on the ground under them, sheltered by the low-growing plants that have sprung up beneath the saplings. But as jack pines continue to grow, they start to block the sunlight that filters down to the forest floor. Deprived of light, the plants growing under the pines begin to die off. Eventually little or no vegetation survives under the pines, and the warblers can no longer nest in the forest.

Clue #1 The habitat of the Kirtland's warbler is forests of jack pine trees.

Clue #2 Kirtland's warblers nest on the ground, under low-growing plants.

Clue #3 In a stand of tall jack pine trees, not much sunlight reaches the forest floor.

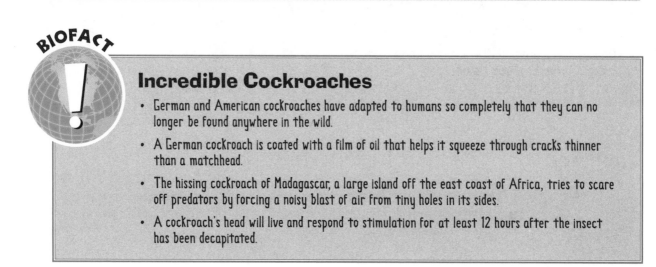

BIOFACT

Incredible Cockroaches

- German and American cockroaches have adapted to humans so completely that they can no longer be found anywhere in the wild.

- A German cockroach is coated with a film of oil that helps it squeeze through cracks thinner than a matchhead.

- The hissing cockroach of Madagascar, a large island off the east coast of Africa, tries to scare off predators by forcing a noisy blast of air from tiny holes in its sides.

- A cockroach's head will live and respond to stimulation for at least 12 hours after the insect has been decapitated.

Mystery 9

A little girl goes to the zoo with her parents. "I wanna see the pink flamingoes!" she says, pulling her mother's arm toward the flamingo exhibit. But when they get to the exhibit, the little girl starts to cry. "What's wrong, sweetie?" asks her father. "You said you wanted to see flamingoes—well, here they are!" The little girl stamps her foot. "But they aren't pink enough!" she says, pouting. The girl's parents look at each other and sigh. "She's got a point," says the mother. "These birds aren't very pink at all. And look at those two over there—they're practically white." Why aren't the flamingoes pink enough?

Answer: Whether or not a flamingo is "in the pink" depends on its diet. In populations of wild flamingoes, eating large quantities of a tiny crustacean causes the birds' feathers to become pink. But in captivity, that same type of crustacean isn't necessarily readily available. Because of this, zoo flamingoes sometimes look like much paler versions of their wild counterparts. But zoo keepers can remedy the situation by supplementing the birds' diet. Sometimes they use the real thing—crustaceans. But other color "enhancers"—including carrot juice—will work in a pinch!

Clue #1 Flamingoes feed on tiny water animals.

Clue #2 In the wild, flamingoes are almost always pink.

Clue #3 Flamingoes living in a zoo often have a different diet from flamingoes living in the wild.

BIOFACT

Forest Under the Sea—Giant kelp is one of the fastest-growing plants on Earth. This seaweed, a kind of algae, may grow as much as 300 feet in 1 year. The vast underwater forests created by these plants are homes to hundreds of different species of animals.

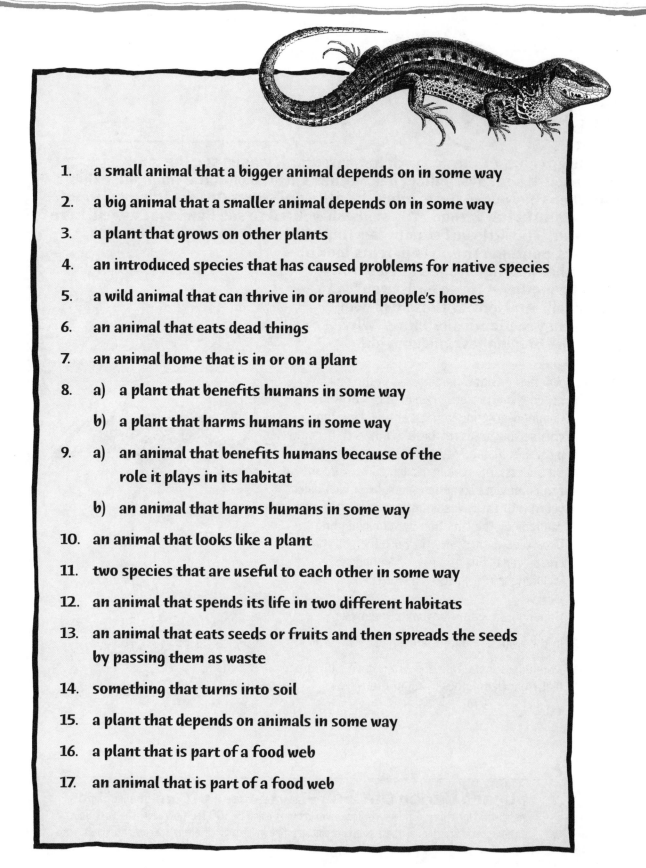

1. a small animal that a bigger animal depends on in some way

2. a big animal that a smaller animal depends on in some way

3. a plant that grows on other plants

4. an introduced species that has caused problems for native species

5. a wild animal that can thrive in or around people's homes

6. an animal that eats dead things

7. an animal home that is in or on a plant

8. a) a plant that benefits humans in some way

    b) a plant that harms humans in some way

9. a) an animal that benefits humans because of the role it plays in its habitat

    b) an animal that harms humans in some way

10. an animal that looks like a plant

11. two species that are useful to each other in some way

12. an animal that spends its life in two different habitats

13. an animal that eats seeds or fruits and then spreads the seeds by passing them as waste

14. something that turns into soil

15. a plant that depends on animals in some way

16. a plant that is part of a food web

17. an animal that is part of a food web

ARTHROPOD PICTURES

Each of the following creatures represents a general group of arthropods. First identify all 10, then write the letter of each one in the correct place on the Arthropod I.D. Chart.

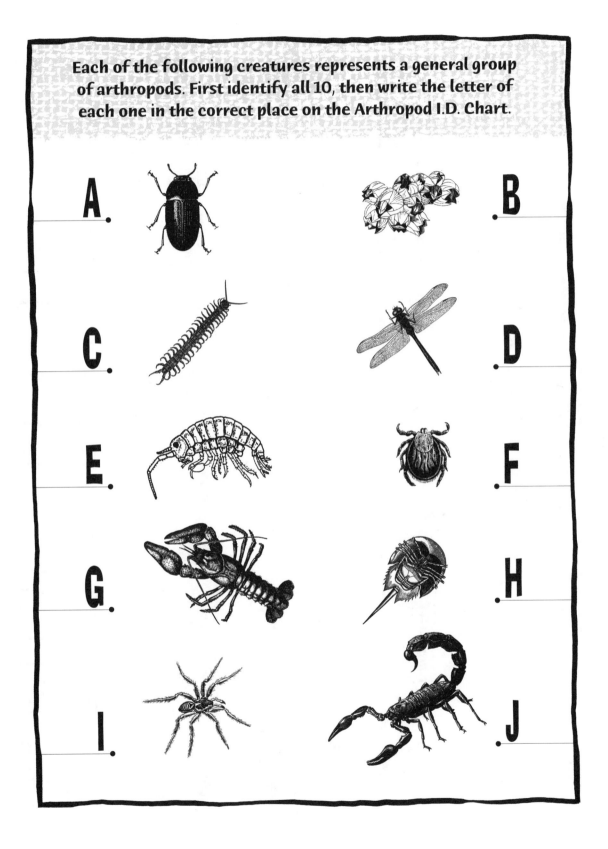

A.

B.

C.

D.

E.

F.

G.

H.

I.

J.

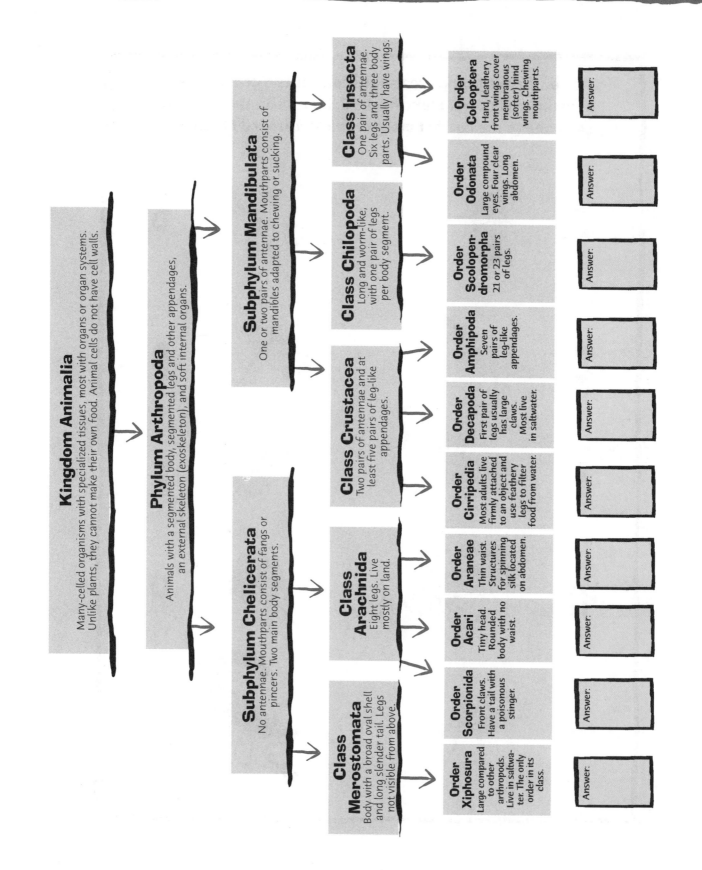

Kingdom Animalia
Many-celled organisms with specialized tissues, most with organs or organ systems. Unlike plants, they cannot make their own food. Animal cells do not have cell walls.

Phylum Arthropoda
Animals with a segmented body, segmented legs and other appendages, an external skeleton (exoskeleton), and soft internal organs.

Subphylum Chelicerata
No antennae. Mouthparts consist of fangs or pincers. Two main body segments.

Subphylum Mandibulata
One or two pairs of antennae. Mouthparts consist of mandibles adapted to chewing or sucking.

Class Merostomata
Body with a broad oval shell and long slender tail. Legs not visible from above.

Class Arachnida
Eight legs. Live mostly on land.

Class Crustacea
Two pairs of antennae and at least five pairs of leg-like appendages.

Class Chilopoda
Long and worm-like, with one pair of legs per body segment.

Class Insecta
One pair of antennae. Six legs and three body parts. Usually have wings.

Order Xiphosura
Large compared to other arthropods. Live in saltwater. The only order in its class.

Order Scorpionida
Front claws. Have a tail with a poisonous stinger.

Order Acari
Tiny head. Rounded body with no waist.

Order Araneae
Thin waist. Structures for spinning silk located on abdomen.

Order Cirripedia
Most adults live firmly attached to an object and use feathery legs to filter food from water.

Order Decapoda
First pair of legs usually has large claws. Most live in saltwater.

Order Amphipoda
Seven pairs of leg-like appendages.

Order Scolopendromorpha
21 or 23 pairs of legs.

Order Odonata
Large compound eyes. Four clear wings. Long abdomen.

Order Coleoptera
Hard, leathery front wings cover membranous (softer) hind wings. Chewing mouthparts.

Answer:

Answer:

Answer:

Answer:

Answer:

Answer:

Answer:

Answer:

Answer:

1. Why isn't there any bacteria growing around the mold?

A Scottish scientist named Alexander Fleming asked this question in 1928. He was studying a type of bacteria that causes throat infections. His laboratory was filled with cultures of bacteria growing in small covered plates called petri dishes. One day, he noticed a fuzzy mold had begun to grow in one of the petri dishes. Fleming was ready to throw the dish away when he noticed something else. Around the fuzzy mold was a clear zone with no bacteria. He wondered why. Fleming predicted (hypothesized) that the mold was producing a substance that caused the bacteria to die. His work on this question led him to isolate the substance produced by the mold. Fleming named this bacteria-killing substance "penicillin" after the mold that produced it.

Fleming's discovery of penicillin-producing mold led to the development of the first antibiotic (a drug that is used to treat infections) and started the search for other antibiotics that could be used for treating bacterial infections.

2. How do migrating animals navigate?

Consider the Arctic tern, a bird that flies as far as 22,000 miles to its wintering grounds and back. Then there are monarch butterflies, some of which fly hundreds of miles to their winter destination. These animals are just two examples of the thousands of species that take on the challenging task of migration each year.

There's a lot that scientists still don't understand about migration, but research is revealing solutions to its mysteries little by little. For example, although scientists have long known that birds and other animals use the Earth's magnetic field to navigate, until recently there wasn't much good evidence to indicate exactly *how* they might sense this field. Now scientists in New Zealand have come up with some intriguing new data suggesting how this might happen. But they didn't find the evidence in a bird or a butterfly. Instead they found it in a fish!

The scientists' research demonstrated that rainbow trout have magnetic particles in their bodies. The research also showed that the fish have a nerve that is influenced by shifting magnetic fields. And in behavioral experiments, the trout themselves responded to magnetic fields. With this work, recently published in the journal *Nature*, scientists are much closer to understanding how animals use their built-in compass.

3. How did the figs get so big and sweet?

Farmers thought that the rich soil and mild climate of California would be perfect for growing figs. But when a large, sweet variety of figs was brought from Smyrna (in southern Turkey), a mystery followed. The trees grew large and healthy—but the figs themselves withered and fell to the ground soon after they started to develop. A horticulturist named George Roeding began working on the problem. First, he went to the area in Turkey where the fig trees had come from. There he found male fig trees with inedible fruit interspersed with fruiting female trees. It turned out that without the pollen from these male trees, the fruits on the female trees couldn't ripen. So Roeding returned to California and tried pollinating the fruit on the female trees using toothpicks. The fruit ripened, but pollinating it took too

fig

much time and effort. He had to go back to Turkey to find out how the fruit got so big and sweet without people to pollinate it.

Roeding discovered that in nature a tiny wasp carries the pollen from the male trees to the female trees, doing all the work that he had done with a toothpick. He brought some of the wasps to California, and they went right to work. And they're still working today! The resulting figs, called Calimyrna (California plus Smyrna), are now a huge hit in the United States.

4. What makes cockleburs stick so tightly?

That's what George DeMestral wanted to know after getting covered with cockleburs during a walk in the Swiss countryside. The cockleburs, which contain seeds from their parent plants, clung to his jacket, just as they cling to fur, feathers, and whatever else they can latch onto. By hitching a ride in this way, the seeds reach new areas in which to sprout and grow.

DeMestral removed the cockleburs from his jacket and looked at them under a microscope. He discovered they were covered with tiny hooks. The hooks are what made the burrs stick to his jacket. DeMestral figured that the same idea could be turned into something very useful. And he was right. The result of DeMestral's fateful walk in the country is Velcro™, a fastening tape product used in everything from athletic shoes to medical supplies.

5. What does the inside of a shark's nose have to do with biodiversity?

That's one of the questions George Benz has been trying to answer for more than 20 years. Dr. Benz is a biologist who began his career studying sharks. He soon realized, though, that you can't really understand sharks without studying the many creatures that live in and on them. For example, if you were to look closely inside the nose of a shark (not that you'd want to), you would find all the normal parts of a nose, along with dozens of tiny creatures hanging onto the slimy tissues inside. You might also find similar

creatures clinging to the shark's fins, scrunched in between its teeth, and even dangling from its eyeballs. What are they? They're copepods—tiny shrimp-like creatures that live pretty much everywhere there's water—in marshes, bogs, rivers, lakes, and oceans.

Thousands of species of copepods can be found throughout the world. Most are scavengers that feed on decaying organisms. But the little freeloaders that Dr. Benz collects are parasites, which means they live off other creatures, getting food and shelter at the creatures' expense. The animals that the parasites live on are called hosts. And some hosts are floating parasite hotels. Dr. Benz has found that a typical blue shark might host 100 copepods on its fins, 4,000 on its gills, and 400 more in its nose. And that's in addition to having 10,000 tapeworms (a type of parasitic worm) squirming around in its small intestine.

To study sharks and their parasites, Dr. Benz travels all over the world. He collects samples from live sharks and other sea creatures, scraping and squeezing to collect the tiny parasites for study back in his lab at the Tennessee Aquarium. There he compares what he finds to what he's already collected and what other biologists have

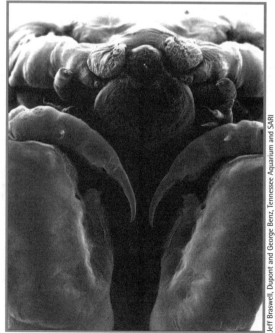

copepod that attaches to the eyes of the Greenland shark

found. The more he learns, the more questions he asks. For example, he and other biologists have noticed that parasites are not just haphazardly scattered on their hosts. Dr. Benz wants to know why. He also wants to know how these creatures find just the right host—and the right spot on that host—to latch onto after swimming in the ocean as tiny larvae. He's also interested in how copepods affect their shark hosts over time and how parasites in general affect shark populations and distribution.

Dr. Benz says we've just scratched the surface of our understanding of parasites—especially when you consider that these creatures might make up more than half the world's species. But maybe if Dr. Benz keeps scraping and squeezing the surfaces of sharks, rays, and other creatures, we'll learn a lot more about marine biodiversity!

6. How do elephants communicate over such long distances?

Scientists who study elephant behavior had been asking that question for years. They had observed that male elephants (bulls) would travel from as far as six miles away when a female was in estrus. It's only during estrus—which occurs for a few days every four years—that female elephants are able to become pregnant.

Scientists wondered how bulls that were so far away were able to find females in estrus, as the females' calls seemed too soft to be heard over such long distances. Then one day in 1985, while working with Indian elephants at Oregon's Portland Zoo, a researcher named Katherine Payne felt a soundless "throbbing" in the air. It reminded her of the sensation she had experienced as a child when she was standing next to a pipe organ as its lowest notes were being played. Payne wondered if the elephants were making infrasonic sounds—sounds too low to be heard by human ears. Her research confirmed that elephants do communicate

infrasonically. Their vocalizations have a sound-pressure equivalent to that of a speeding train and can be heard by elephants miles away.

Researchers are now trying to decipher the infrasonic language of elephants. They're also studying the effects of factors such as wind speed, air temperature, and thunderstorms on elephant communication.

7. Why did the apple fall to the ground?

That was the question that occurred to Isaac Newton one summer day in the 1660s. Sitting in his garden under an apple tree, Newton was in a thoughtful mood. When he saw an apple fall from the tree—as he no doubt had seen many times before—something seemed to click in his curious brain. Why, he wondered, did the apple fall to the ground instead of flying off sideways or even drifting off into space? At that moment Newton came up with a possible answer that was to change science forever. He decided that the Earth itself must somehow have "attracted" the apple.

In this way Newton discovered the concept of gravity. But his momentous discovery didn't end with the apple. He figured that gravity was a force found throughout the universe, and he used his ideas about this force to consider, among other things, the distances between planets and how the moon stays in orbit around the Earth.

How much do you know about where you live?

1. What major habitat type do you live in? (temperate forest, temperate rain forest, grassland, shrubland, taiga, tundra, desert, and so on)

2. Name three native trees that live in your area. Collect a leaf from each one.

3. Name five native edible plants that grow in your region, and list in which season(s) each is available.

4. Name one poisonous plant that lives in your area.

5. Name ten native animals that live in your region.

6. Name three native animals that you can see in your area at any time of the year.

7. Name three migratory animals that live in your area, and list in which season(s) you're able to see them.

8. Do deer live in your area? If so, when during the year do they give birth?

9. How much average rainfall does your community get each year?

10. When (during what season or month) does your community normally get the most precipitation?

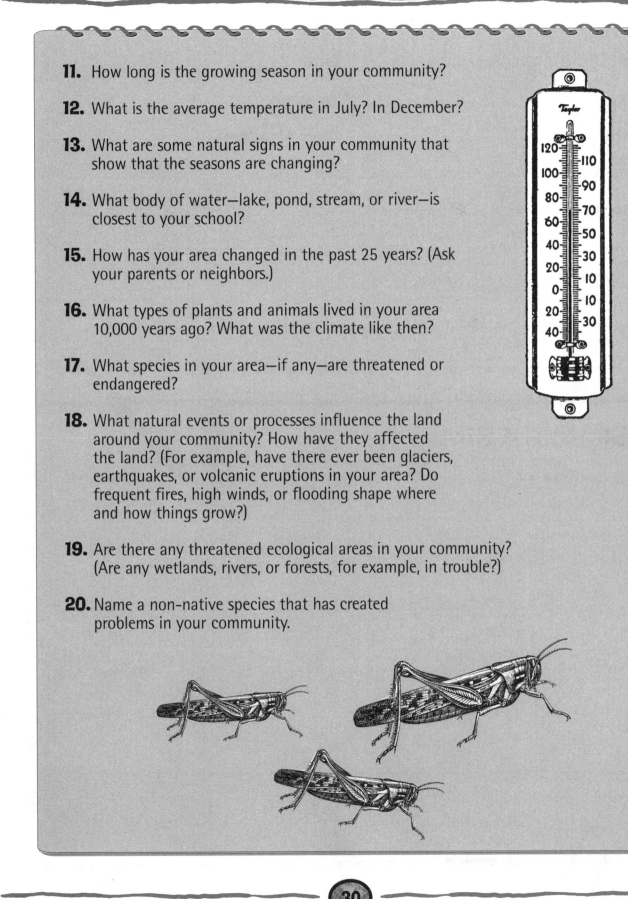

11. How long is the growing season in your community?

12. What is the average temperature in July? In December?

13. What are some natural signs in your community that show that the seasons are changing?

14. What body of water—lake, pond, stream, or river—is closest to your school?

15. How has your area changed in the past 25 years? (Ask your parents or neighbors.)

16. What types of plants and animals lived in your area 10,000 years ago? What was the climate like then?

17. What species in your area—if any—are threatened or endangered?

18. What natural events or processes influence the land around your community? How have they affected the land? (For example, have there ever been glaciers, earthquakes, or volcanic eruptions in your area? Do frequent fires, high winds, or flooding shape where and how things grow?)

19. Are there any threatened ecological areas in your community? (Are any wetlands, rivers, or forests, for example, in trouble?)

20. Name a non-native species that has created problems in your community.

Site ....

City ....

State ....

Temperature ....

Description *(what the area looks like in general)*

....

....

....

....

Date ....

Weather ....

Team Members ....

....

....

Sketch of Site:

Plants

....
....
....
....
....
....
....
....
....

Insects

....
....
....
....
....
....
....
....
....
....

Non-insect Invertebrates

....
....
....
....
....
....
....

Mammals

....
....
....
....
....
....
....

Birds

....
....
....
....
....
....
....

Reptiles and Amphibians

....
....
....
....
....
....

Other

....
....
....
....
....
....

Is It an Insect?

It may seem difficult to tell the difference between a true insect and some of the other little creatures you find scurrying around. But if you look out for these two characteristics— things that only insects share—you'll find it's not so hard.

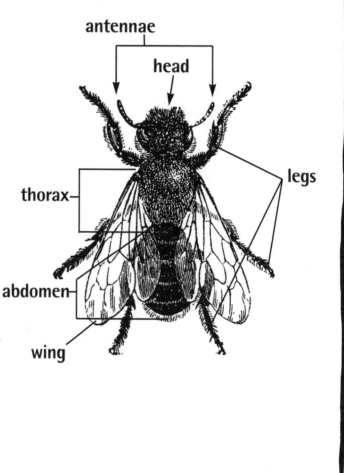

antennae

head

thorax

abdomen

wing

legs

1. Insects have three main body parts: a **head**, a **thorax**, and an **abdomen**.

2. Adult insects have **six legs** (three pairs of legs).

Another hint–If you find any tiny creature that has **wings,** it's definitely an insect. Insects are the only small creatures that have mastered the art of true flight. Some may have two wings and others may have four. And in some groups, one set of wings can act as a cover for the other set. Just remember: Not all insects have wings.

Scientists put insects that are similar to each other into groups called **orders**. There are about 30 main orders of insects. The chart titled "Putting Them in Order" will help you identify insects from some of the most common orders you're likely to sweep up.

One Strong Beetle—Ounce for ounce, the strongest animal on Earth may be the rhinoceros beetle. In an experiment, this inch-long insect carried 30 times its own weight on its back for a distance that would be equal to a person walking a mile carrying a pickup truck.

Scientists put insects that are similar to each other into groups called orders. There are about 30 main orders of insects. This chart will help you identify insects from some of the most common orders you're likely to sweep up.

Picture	Common Name	Order	Description
	springtails	Collembola	are tiny, jumping creatures that live in soil, decaying logs, and leaf mold (they jump by releasing a forked structure on their abdomen); color varies from white to red to mottled; hard to see because of small size
	mayflies	Ephemeroptera	have a long soft body; have two or three thread-like tails; are very common around ponds or streams; front wings are large and triangular and hind wings are small and rounded
	damselflies dragonflies	Odonata	have four wings with many veins and are erratic fliers; have large compound eyes; are often brightly colored and found around water; length varies from about 1 to 5 inches; young live in water
	grasshoppers crickets katydids	Orthoptera	often have four long and narrow wings; hind wings fold under leathery front wings; color varies; can make sounds by rubbing one body part against another; large hind legs are used for jumping
	walking sticks	Phasmatodea	look like twigs or sticks; have brownish or green bodies; have long slender legs and move slowly; usually found on trees and shrubs; most are wingless in the United States

	cockroaches	Blattodea	have an oval, flattened body and long, hair-like antennae; have slender front and hind legs and are often fast runners; some have wings, but others are wingless; in some species the hind wings fold under leathery front wings
	earwigs	Dermaptera	have long, slender bodies with a pincer-like structure, called a cerci, on their abdomen; adults usually have four wings; when at rest, the membranous hind wings fold under the short and leathery front wings
	termites	Isoptera	are small to medium-sized insects that live in social groups; front and hind wings (if present) are same size and held flat over the body; workers are pale and wingless; reproductive kings and queens have wings and compound eyes and swarm during mating season
	stoneflies	Plecoptera	have four wings with many veins; are poor fliers; are small or medium-sized and brown or drab-colored; are usually found near water; young live in water
	thrips	Thysanoptera	are very small, slender-bodied insects; some have four long, narrow wings fringed with long hairs; some are wingless; many are plant feeders and are often found in the flowerheads of daisies and dandelions
	ambush bugs assassin bugs stink bugs	Hemiptera	body is broad or long and narrow; front wings are half leathery and half membranous and make a triangle where they fold across the abdomen; live in almost all habitats and have piercing-sucking mouthparts
	cicadas aphids leafhoppers	Homoptera	are closely related to the true bugs; many have four wings, although some, such as scale insects, don't have wings; wings at rest are held roof-like over the body; antennae are often short and bristle-like

	dobsonflies lacewings alderflies	Neuroptera	are soft-bodied insects with four membranous wings that have many veins; wings are held roof-like over the body when at rest; antennae are usually long and have many segments; adults are usually weak fliers
	beetles	Coleoptera	hind wings fold beneath hardened front wings; front wings make a line straight down their back where they meet; have chewing mouthparts; antennae come in a variety of shapes
	butterflies moths	Lepidoptera	have four wings, which are covered with scales that come off like dust when handled; have sucking mouthparts in the form of a coiled tube; butterflies usually have club-like antennae
	flies (mosquitoes, house flies, and so on)	Diptera	are usually small and soft-bodied; have two clear front wings; hind wings reduced to two tiny knobbed structures called halteres that help flies keep their balance while flying
	fleas	Siphonaptera	are small, wingless insects that feed on the blood of birds and mammals; have flattened bodies and strong jumping legs; adults have sucking mouthparts and short antennae; body is covered with bristles and spines
	ants bees wasps	Hymenoptera	have four clear wings; abdomen usually is narrowly attached to thorax by a thin "waist"; often have a stinger at the tip of the abdomen

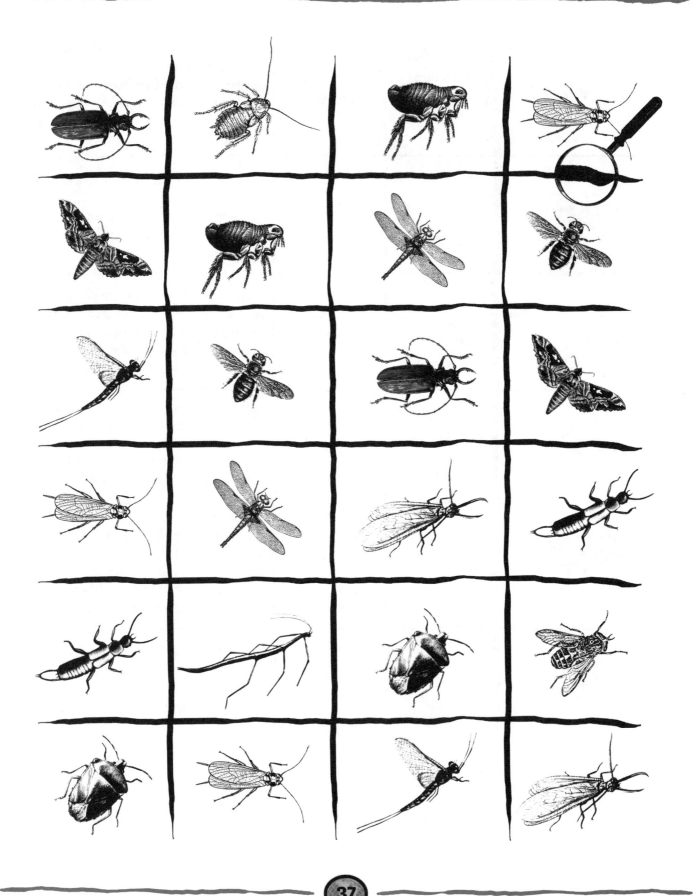

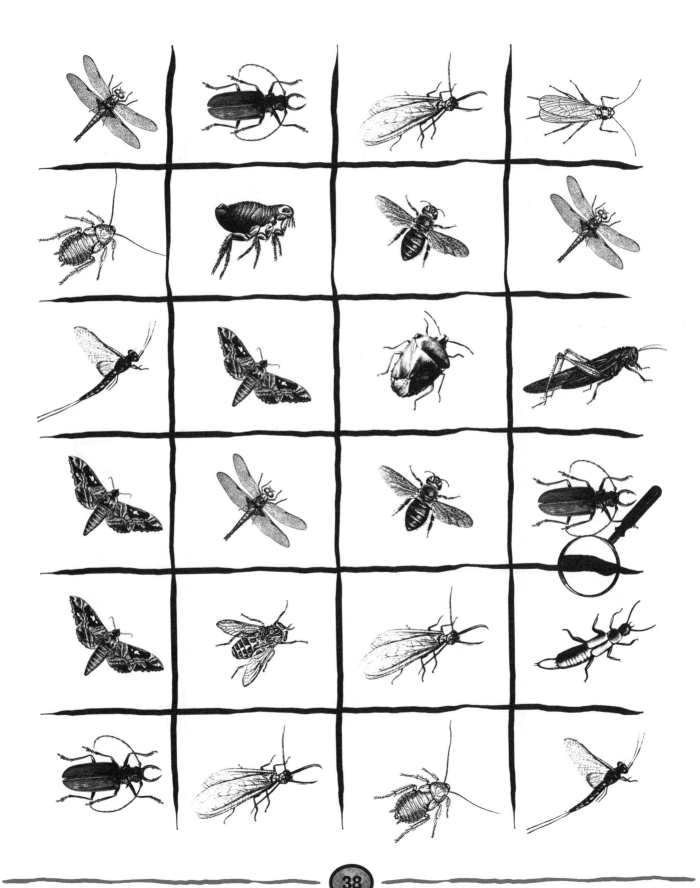

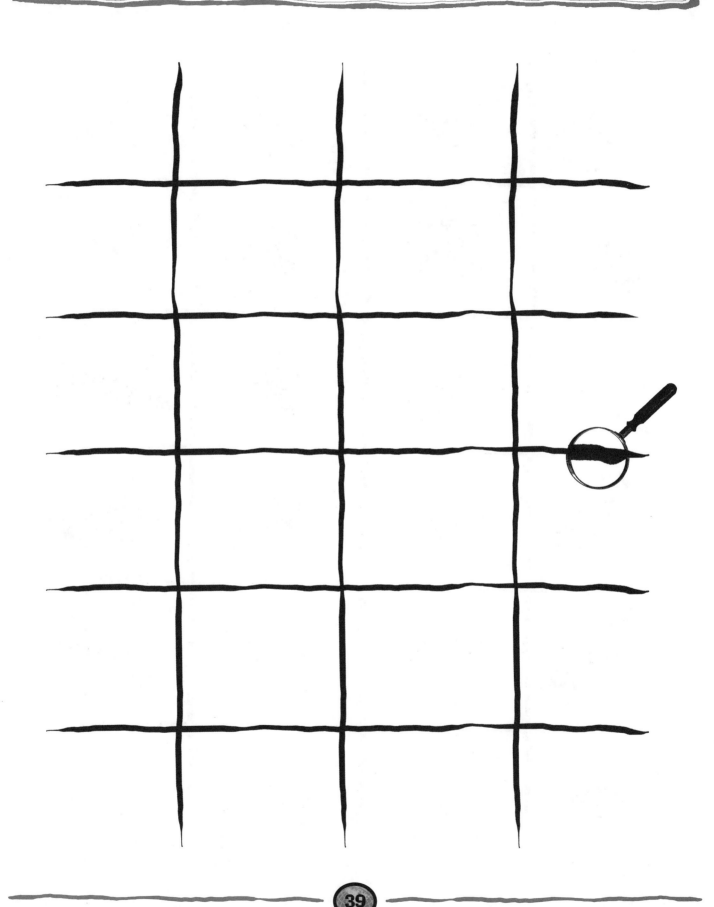

Which of the following traits did you inherit from your parents?
Check the box next to the trait that best describes you.

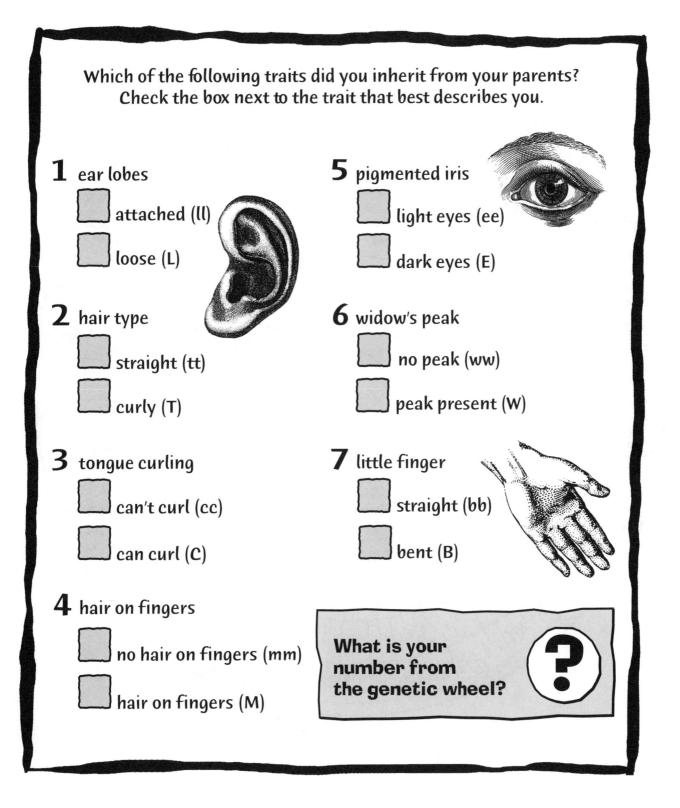

1 ear lobes

☐ attached (ll)

☐ loose (L)

2 hair type

☐ straight (tt)

☐ curly (T)

3 tongue curling

☐ can't curl (cc)

☐ can curl (C)

4 hair on fingers

☐ no hair on fingers (mm)

☐ hair on fingers (M)

5 pigmented iris

☐ light eyes (ee)

☐ dark eyes (E)

6 widow's peak

☐ no peak (ww)

☐ peak present (W)

7 little finger

☐ straight (bb)

☐ bent (B)

What is your number from the genetic wheel? **?**

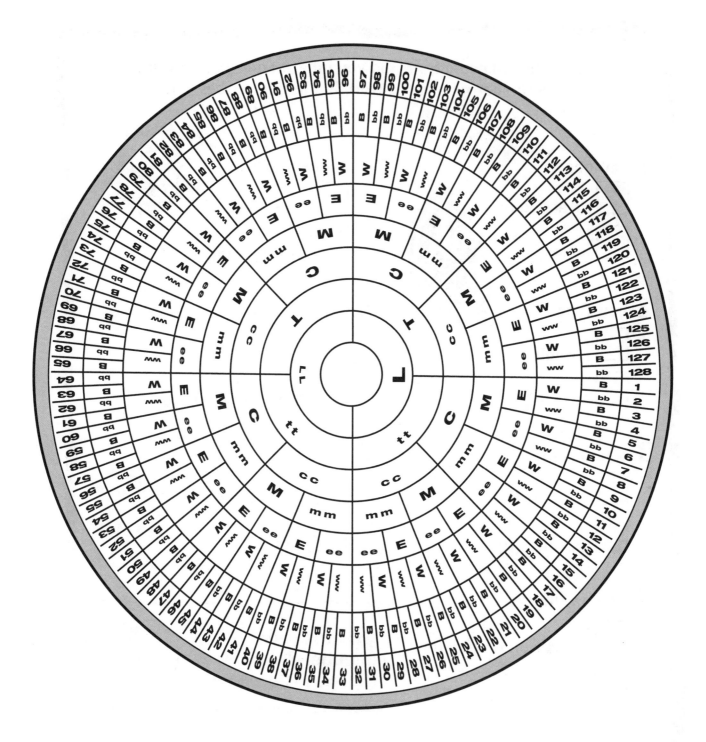

Giraffes really know how to draw a crowd! On a typical scorching day on the African plains, these gentle giants are surrounded by an amazing variety of grazers. Ostriches, kudu, impalas, wildebeests, and gazelles all gather near. No wonder—with their keen eyesight and unique vantage point, giraffes can easily spot a moving lion, hyena, or human predator from a mile away.

These sentinels of the savanna are certainly well known for their interesting bodies. Their necks are the longest of any living mammal, although they have the same number of vertebrae in their necks as humans do.

Giraffes are the ultimate tree browsers, their bodies providing marvelous evidence of the efficient design of nature. To begin with, the giraffe's long neck and enormous vertebrae allow it to browse on tree leaves that few other animals can reach. To get blood all the way to the animal's head requires a huge heart. The average giraffe's heart is 2 feet long and weighs 25 pounds, compared to a human heart that is only 4 to 5 inches long and weighs 1 pound. A special hinge at the base of its head allows a giraffe to hold its long slender head in a straight line with its neck, giving it two more feet of reach. Add to that a prehensile upper lip that can grasp like fingers and a tongue that can reach 18 inches farther, and you have a browser that's hard to beat.

Giraffes can reach leaves that other browsers can't, and their favorite food is the acacia tree—a tree that most animals wouldn't be able to eat, even if they could reach it. Acacia leaves are highly nutritious and moist, giving giraffes almost all the nutrients and much of the water they need. But acacia branches are covered with thick thorns. Giraffes have strong hairs and thick skin, which protect their faces from these sharp thorns. To top it all off, the giraffe's eyes are larger than those of any other land mammal, and they are situated on the sides of its head, enabling the giraffe to be alert to danger at any moment. Giraffes are able to keep watchful guard across the plains, as they never sleep for more than four to five minutes at a time.

Although they have many helpful adaptations, there is still no such thing as the perfect giraffe. Their environment is always changing, undergoing seasonal fluctuations as well as long-term climate changes. So, diversity within giraffe populations has been the key to the animal's survival and evolution. The variation among giraffes is so great that there are nine different subspecies, all distinguishable by their spot patterns, size, and number of horns. For example, Nubian giraffes have dark spots with irregular edges, Transvaal giraffes have spots with finger-like projections, and reticulated giraffes have a more regular, net-like pattern of dark patches divided by light lines.

Giraffes may occur singly, but they usually live in herds of up to 100 or more, with the individuals loosely associated. The females and calves graze in family groups and are often far apart from other group members. They change groups frequently while searching for

42

food. Giraffes are so tall that they can see other members of their herd from long distances. While grazing, giraffe calves and mothers are sometimes separated from each other by more than a mile and for several days at a time. The males move between herds, browsing and searching for mates. This constant flow and intermingling of giraffes within the larger population is important for maintaining genetic diversity. And because all the subspecies interbreed, there is an amazing variety of giraffes in the total population, so much so that no two giraffes have the same pattern of spots! This fact has enabled scientists to use the spot patterns of giraffe necks to distinguish individuals.

Although giraffes are not endangered, there is no doubt that their current populations are much smaller than they used to be. At one time, giraffes ranged all over Africa as well as southern Europe and southeastern Asia. Over thousands of years, changes in climate gradually forced the ancestors of modern giraffes to either migrate or die out. For example, the climate changes that led to the formation of the Sahara Desert a few thousand years ago eventually forced giraffe populations southward. Today the only places where you can find giraffes in the wild are along a narrow belt across the center of Africa and in a small section of southern Africa.

Despite their apparently healthy numbers, giraffes are certainly not immune to danger. Their biggest problem is the same as that of most African animals: lack of space. Expanding human populations mean shrinking habitats for wildlife, posing particular problems for large, migratory animals such as elephants and

giraffes. During one 24-hour period, a large male giraffe can consume 75 pounds of food and may wander up to 20 miles in search of food and water. This wandering can become risky when giraffes leave the safety of national parks and protected areas.

It's no wonder that giraffes rarely allow people to get closer than 100 feet. Although lions and hyenas attack giraffe calves, humans are the only predator of adults. People throughout history have hunted giraffes for a variety of reasons. Although African governments have passed laws to make sport hunting illegal, giraffes are still killed for their meat, hides, and tails. Giraffe skins are used to make shoes, harnesses, whips, and shields. And their tails are popular for making amulets and bracelets. Some giraffes are killed simply because they wander into farmland or knock down power lines.

The future of giraffes is tied to the future of the people living around and among them. Some farmers have discovered that, by allowing giraffes to graze on the acacia trees on their farms or pastures, the giraffes actually perform a service by pruning the trees and keeping them from overgrowing. Giraffes are also a huge draw for tourists, who bring much-needed income to many African countries.

As you play the simulation game, it is important to keep in mind that the events described are fictional scenarios. They are, however, based on real events that face giraffes and other wildlife in Africa and around the world. Thinking about these events and how they might affect wild populations will help us to appreciate the importance of preserving genetic diversity within a species.

43

Leafy-spotted

Reticulated

GIRAFFE GENETIC WHEEL

BEGIN WITH THE CENTER CIRCLE AND MOVE OUTWARD, BASED ON YOUR GIRAFFE'S TRAITS.

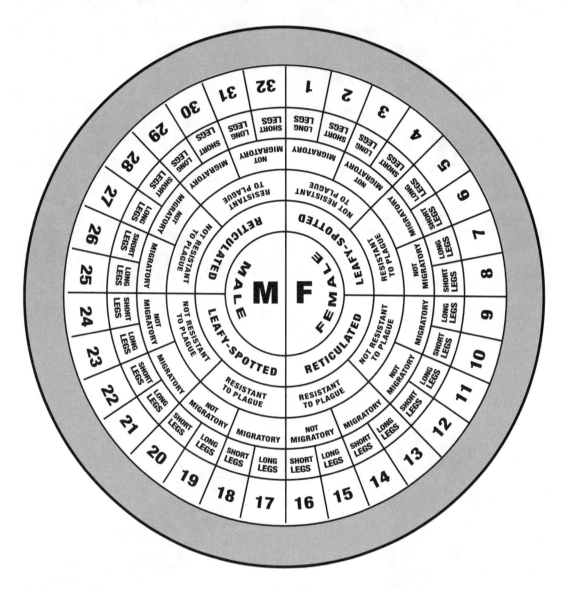

For example, a female giraffe with the following characteristics:

Reticulated

Not Resistant to Plague

Not Migratory

Long Legs

would have a genetic number of **11**.

A male with the same characteristics would have a genetic number of **27**.

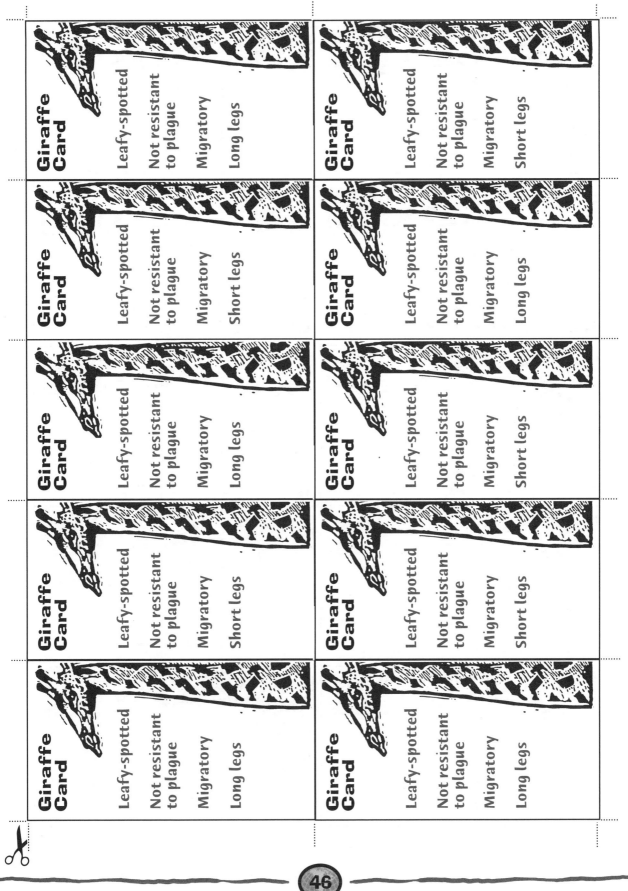

Giraffe Card
Leafy-spotted
Not resistant to plague
Migratory
Long legs

Giraffe Card
Leafy-spotted
Not resistant to plague
Not migratory
Long legs

Giraffe Card
Leafy-spotted
Not resistant to plague
Migratory
Short legs

Giraffe Card
Leafy-spotted
Not resistant to plague
Migratory
Short legs

Giraffe Card
Leafy-spotted
Not resistant to plague
Migratory
Long legs

Giraffe Card
Leafy-spotted
Not resistant to plague
Not migratory
Long legs

Giraffe Card
Leafy-spotted
Not resistant to plague
Migratory
Short legs

Giraffe Card
Leafy-spotted
Not resistant to plague
Migratory
Short legs

Giraffe Card
Leafy-spotted
Resistant to plague
Migratory
Short legs

Giraffe Card
Leafy-spotted
Not resistant to plague
Not migratory
Long legs

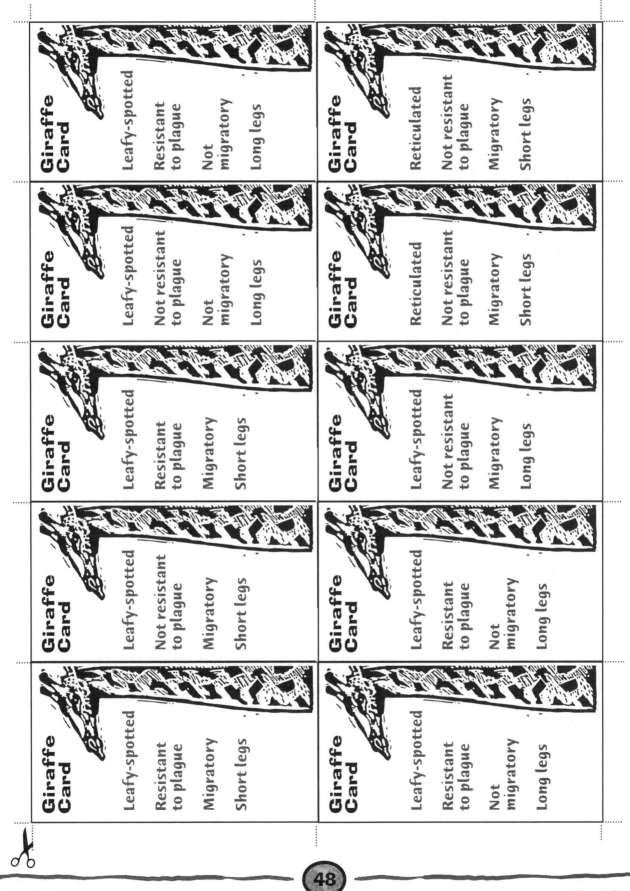

Giraffe Card
- Leafy-spotted
- Resistant to plague
- Not migratory
- Long legs

Giraffe Card
- Reticulated
- Not resistant to plague
- Migratory
- Short legs

Giraffe Card
- Leafy-spotted
- Not resistant to plague
- Not migratory
- Long legs

Giraffe Card
- Reticulated
- Not resistant to plague
- Migratory
- Short legs

Giraffe Card
- Leafy-spotted
- Resistant to plague
- Migratory
- Short legs

Giraffe Card
- Leafy-spotted
- Not resistant to plague
- Migratory
- Long legs

Giraffe Card
- Leafy-spotted
- Not resistant to plague
- Migratory
- Short legs

Giraffe Card
- Leafy-spotted
- Resistant to plague
- Not migratory
- Long legs

Giraffe Card
- Leafy-spotted
- Resistant to plague
- Migratory
- Short legs

Giraffe Card
- Leafy-spotted
- Resistant to plague
- Not migratory
- Long legs

What Is Biodiversity?

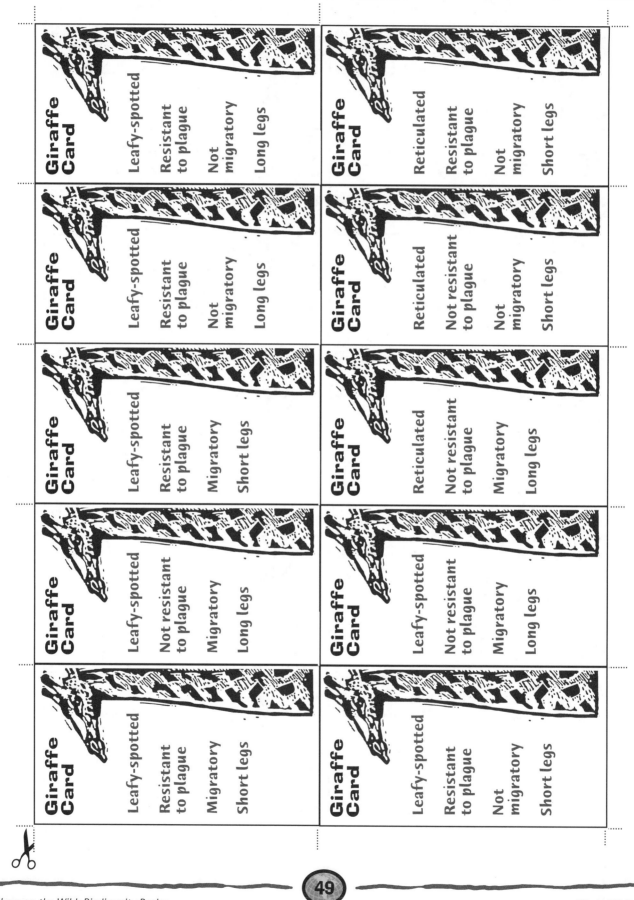

Giraffe Card
- Leafy-spotted
- Resistant to plague
- Not migratory
- Long legs

Giraffe Card
- Reticulated
- Resistant to plague
- Not migratory
- Short legs

Giraffe Card
- Leafy-spotted
- Resistant to plague
- Not migratory
- Long legs

Giraffe Card
- Reticulated
- Not resistant to plague
- Not migratory
- Short legs

Giraffe Card
- Leafy-spotted
- Resistant to plague
- Migratory
- Short legs

Giraffe Card
- Reticulated
- Not resistant to plague
- Migratory
- Long legs

Giraffe Card
- Leafy-spotted
- Not resistant to plague
- Migratory
- Long legs

Giraffe Card
- Leafy-spotted
- Not resistant to plague
- Migratory
- Long legs

Giraffe Card
- Leafy-spotted
- Resistant to plague
- Migratory
- Short legs

Giraffe Card
- Leafy-spotted
- Resistant to plague
- Not migratory
- Short legs

Giraffe Card
Reticulated
Resistant to plague
Not migratory
Long legs

Giraffe Card
Reticulated
Resistant to plague
Not migratory
Short legs

Giraffe Card
Reticulated
Resistant to plague
Migratory
Long legs

Giraffe Card
Reticulated
Resistant to plague
Migratory
Short legs

Giraffe Card
Reticulated
Not resistant to plague
Migratory
Long legs

Giraffe Card
Reticulated
Not resistant to plague
Migratory
Short legs

Giraffe Card
Leafy-spotted
Not resistant to plague
Not migratory
Long legs

Giraffe Card
Leafy-spotted
Not resistant to plague
Not migratory
Short legs

Giraffe Card
Leafy-spotted
Resistant to plague
Migratory
Long legs

Giraffe Card
Leafy-spotted
Resistant to plague
Not migratory
Short legs

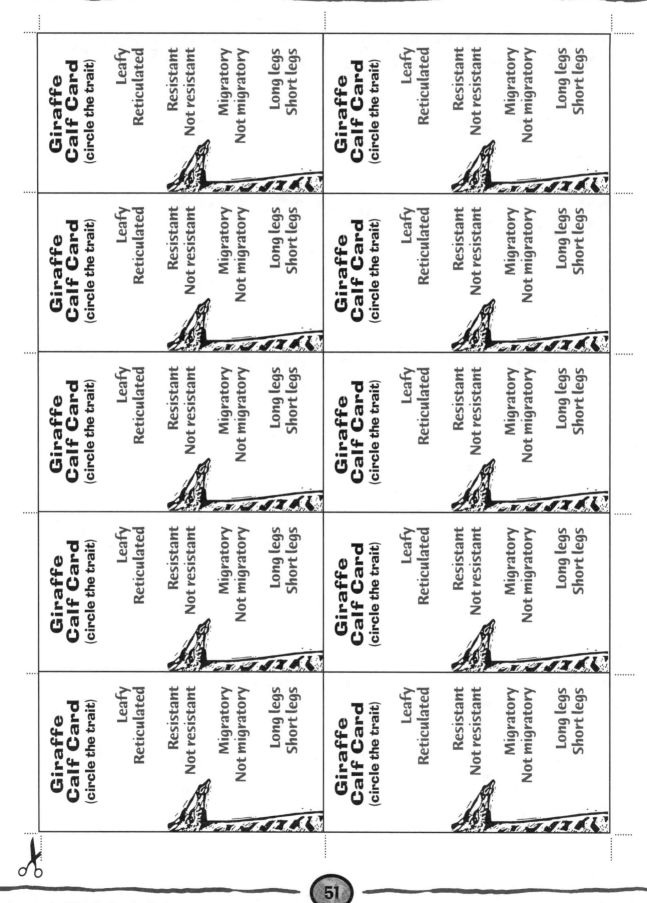

The following text appears on each of the ten cards:

Giraffe Calf Card (circle the trait)

Leafy / Reticulated

Resistant / Not resistant

Migratory / Not migratory

Long legs / Short legs

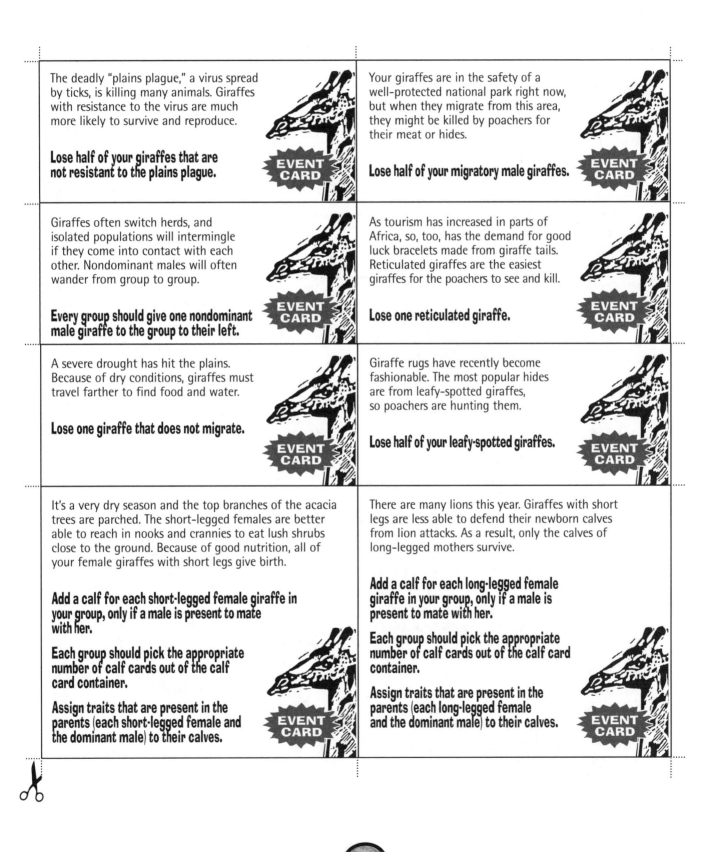

The deadly "plains plague," a virus spread by ticks, is killing many animals. Giraffes with resistance to the virus are much more likely to survive and reproduce.

Lose half of your giraffes that are not resistant to the plains plague.

EVENT CARD

Your giraffes are in the safety of a well-protected national park right now, but when they migrate from this area, they might be killed by poachers for their meat or hides.

Lose half of your migratory male giraffes.

EVENT CARD

Giraffes often switch herds, and isolated populations will intermingle if they come into contact with each other. Nondominant males will often wander from group to group.

Every group should give one nondominant male giraffe to the group to their left.

EVENT CARD

As tourism has increased in parts of Africa, so, too, has the demand for good luck bracelets made from giraffe tails. Reticulated giraffes are the easiest giraffes for the poachers to see and kill.

Lose one reticulated giraffe.

EVENT CARD

A severe drought has hit the plains. Because of dry conditions, giraffes must travel farther to find food and water.

Lose one giraffe that does not migrate.

EVENT CARD

Giraffe rugs have recently become fashionable. The most popular hides are from leafy-spotted giraffes, so poachers are hunting them.

Lose half of your leafy-spotted giraffes.

EVENT CARD

It's a very dry season and the top branches of the acacia trees are parched. The short-legged females are better able to reach in nooks and crannies to eat lush shrubs close to the ground. Because of good nutrition, all of your female giraffes with short legs give birth.

Add a calf for each short-legged female giraffe in your group, only if a male is present to mate with her.

Each group should pick the appropriate number of calf cards out of the calf card container.

Assign traits that are present in the parents (each short-legged female and the dominant male) to their calves.

EVENT CARD

There are many lions this year. Giraffes with short legs are less able to defend their newborn calves from lion attacks. As a result, only the calves of long-legged mothers survive.

Add a calf for each long-legged female giraffe in your group, only if a male is present to mate with her.

Each group should pick the appropriate number of calf cards out of the calf card container.

Assign traits that are present in the parents (each long-legged female and the dominant male) to their calves.

EVENT CARD

September 8, 1885

Yesterday morning we arrived amidst great winds and rains to an uninhabited island. I have named it Sandy Island because of its great sandy dunes. Our first night passed safely, and we awoke to clear skies and warmer temperatures. Immediately, I began to explore it to see what manner of plants and animals occupy its corners. My group and I have become well acquainted with a small reptile that skitters over the beach, digging with a shovel-shaped snout. I imagine it is digging for insects or tiny sea creatures left by the tides. My 12-year-old nephew, James, has dubbed this animal the twitomite, and the name has stuck. He's very excited about the discovery and has sketched it for me on these journal pages.

shovel-shaped snout used for scooping up tiny creatures from the sand

Shovel-Nosed Twitomite

September 16

We sailed about six miles southwest of Sandy Island to arrive at a set of dry, desert-like islands. The first island we visited we called Cactus Island because the beach quickly gives way to an arid landscape of low shrubs and cacti. Here we found another twitomite! It looks a bit like the shovel-nosed twitomite, but it has a longer snout and saw-shaped teeth. We've seen it use these teeth to cut into the many fruits that grow on the cacti. There must be food and water alike in these fruits that help the saw-toothed twitomite live.

We spotted another twitomite on an island we called Flora Island in honor of its great blooming desert plants. James was the first to spot this twitomite's unique feature—it has a long snout with a long, curled tongue it uses to probe the inside

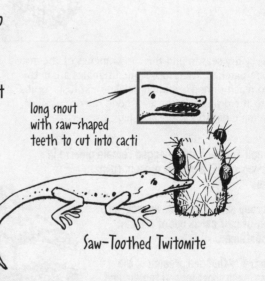

long snout with saw-shaped teeth to cut into cacti

Saw-Toothed Twitomite

of the plants' tubular pink flowers. Curious indeed, I plucked one of these flowers and cut it open. I discovered the inside of the flower is full of a sweet nectar. This nectar seems to attract insects, but it is so thick and sticky that not all of them escape. Thus the long-tongued twitomite can slip its tongue inside the flower and lap up a tasty meal of sweet plant nectar and dead insects all at once.

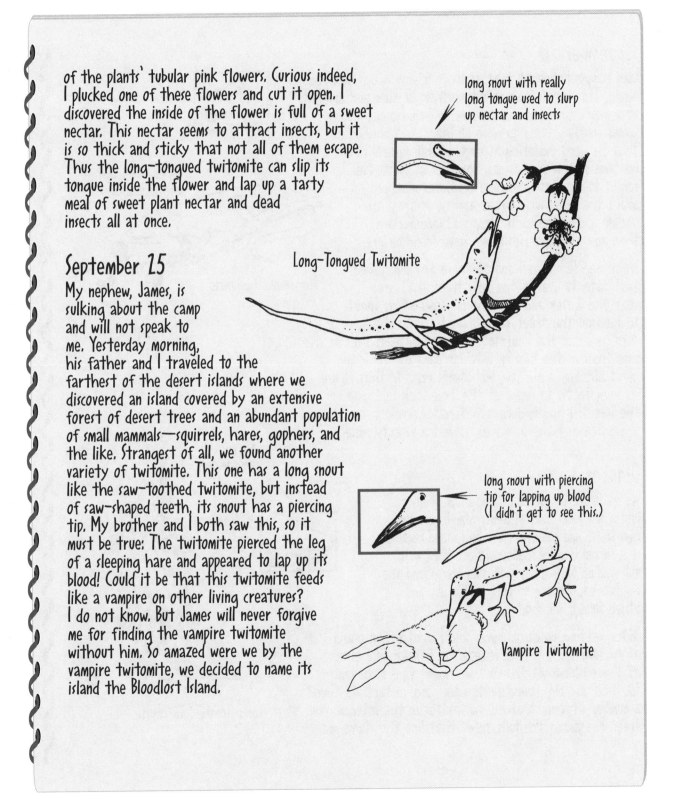

long snout with really long tongue used to slurp up nectar and insects

Long-Tongued Twitomite

September 25

My nephew, James, is sulking about the camp and will not speak to me. Yesterday morning, his father and I traveled to the farthest of the desert islands where we discovered an island covered by an extensive forest of desert trees and an abundant population of small mammals—squirrels, hares, gophers, and the like. Strangest of all, we found another variety of twitomite. This one has a long snout like the saw-toothed twitomite, but instead of saw-shaped teeth, its snout has a piercing tip. My brother and I both saw this, so it must be true: The twitomite pierced the leg of a sleeping hare and appeared to lap up its blood! Could it be that this twitomite feeds like a vampire on other living creatures? I do not know. But James will never forgive me for finding the vampire twitomite without him. So amazed were we by the vampire twitomite, we decided to name its island the Bloodlost Island.

long snout with piercing tip for lapping up blood (I didn't get to see this.)

Vampire Twitomite

September 28

I am happy to report that James is feeling good-humored again. This morning we sailed another 10 miles and arrived at a moist, forested island we've dubbed James Island. James was so pleased to have an island as a namesake that he went exploring at once. Within several hours, he came back with two new varieties of twitomites for us to see. Both of these creatures scamper about the trees with great agility, as they are smaller and have shorter legs and longer claws than any of the varieties we have found before.

We've decided to call one of these the big-jawed twitomite. It lives in the lush trees that grow near the water and uses the big jaws of its snout to feed on the trees' large, sweet fruits. The other we call the long-toothed twitomite. It lives in the pine trees in the higher, interior part of the island. Its snout also has a big jaw, but James reports that it uses its long teeth to gnaw at the tree bark and snap up the insects living underneath. What marvelous tools these lizards have developed to get a hearty meal!

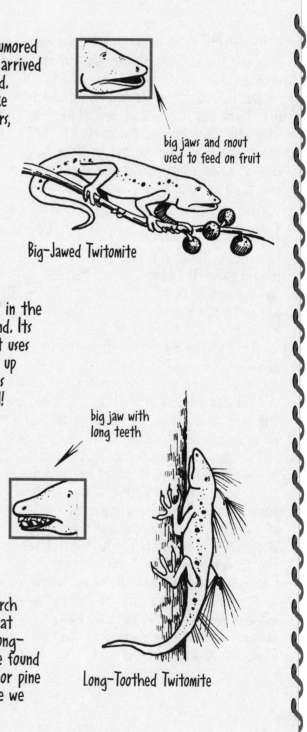

big jaws and snout used to feed on fruit

Big-Jawed Twitomite

big jaw with long teeth

Long-Toothed Twitomite

October 1

We made one of our last island stops today— just a few miles east of James Island. Everyone says we should call this place Bird Island because it is surrounded by shallow waters filled with tall wading birds. They walk and fly around the beach area, plucking up fish, small crabs, and other animals for food.

After setting up camp, James and I went off in search of twitomites. Since the vegetation is much like that of James Island, we thought we might find both long-toothed and big-jawed twitomites. And, in fact, we found a number of long-toothed twitomites in the interior pine trees. But along the lush trees that line the shore we found something truly amazing—a twitomite that looks much like the big-jawed twitomite except that it is green. This twitomite happily ate the fruits of the lush trees while the many birds hunted for their meals. James thinks there must be a special reason the green twitomite came to be, but we can't imagine what it is.

has green skin

Green Twitomite

October 2

Since the winter storms have started to set in, we returned to Sandy Island to prepare for the journey home. James wanted to explore just one more new place. So we took advantage of the good weather this morning and rowed out to a rocky island a half a mile from Sandy Island. As we turned over rocks and collected shells in the tidal pools along the rocky shore, again we came upon an amazing discovery— yet another variety of twitomite. This one looks much like the shovel-nosed twitomite except that its snout is small and pointed, and this variety has webbed feet that allow it to swim through the water. And it can dive and swim— undoubtedly eating the small sea creatures trapped in the shallow pools that collect among the rocks as the tide falls. We have called it the marine twitomite.

Marine Twitomite

October 3

We are sailing home from our newfound islands, sad to leave them and their fascinating inhabitants behind. We still wonder why there are so many different kinds of twitomites on these islands, each with its own particular shape and habits. James says it's a puzzle he's determined to figure out. And perhaps someday he, or someone else, will!

webbed feet for swimming

DIRECTIONS—As you read the journals and look at the sketches, fill in this chart to get a better picture of twitomite diversity.

	Snout	Island/Habitat	Food Source	Outstanding Trait	How does this trait help it to survive?
Shovel-Nosed Twitomite	shovel-shaped	Sandy Island beach	insects and sea creatures	shovel-shaped snout	The shovel-shaped snout allows it to easily dig through the sand and eat insects and sea creatures.
Saw-Toothed Twitomite					
Long-Tongued Twitomite					
Vampire Twitomite					
Big-Jawed Twitomite					
Long-Toothed Twitomite					
Green Twitomite					
Marine Twitomite					

1. How many species of twitomites were there 20,000 years ago? Were the individual twitomites genetically identical? Explain.

2. When did the twitomites first become separated into three distinct populations? Do you think the twitomite populations on the three islands were fairly similar or radically different at first?

3. Over time, what could have caused different species of twitomites to arise on each of the three islands? Be specific.

4. What had happened to the islands by the time the explorers arrived? How did this affect the twitomites?

5. Explain how each of the eight twitomites is particularly well adapted to its habitat.

6. Some scientists think about the process of speciation in terms of this formula:

isolation + time + evolution = SPECIATION

Apply this formula to twitomites to explain how their single ancestor evolved into eight different species.

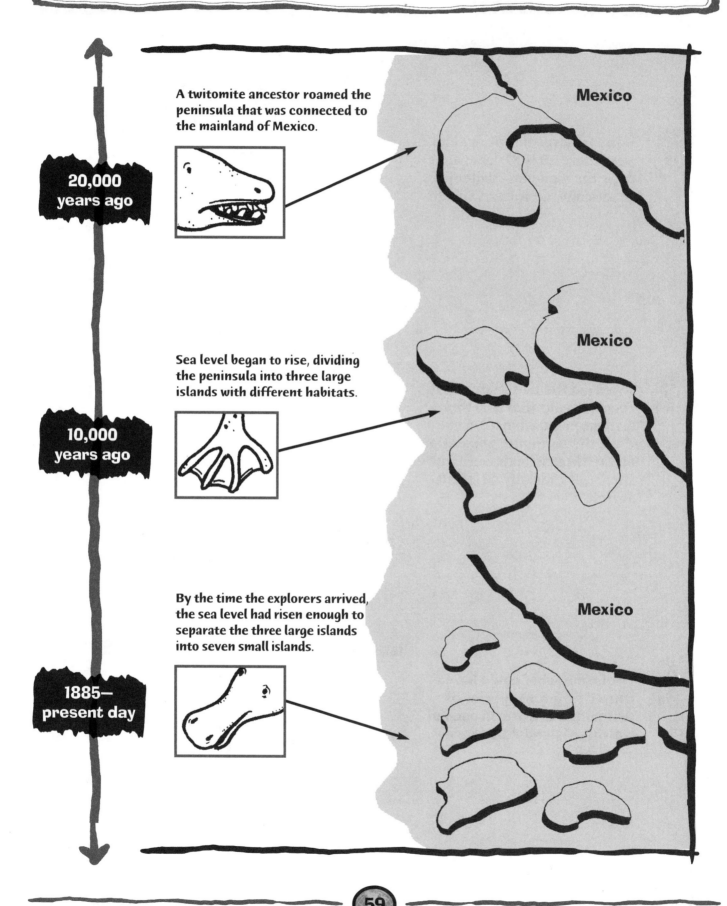

20,000 years ago

A twitomite ancestor roamed the peninsula that was connected to the mainland of Mexico.

Mexico

10,000 years ago

Sea level began to rise, dividing the peninsula into three large islands with different habitats.

Mexico

1885— present day

By the time the explorers arrived, the sea level had risen enough to separate the three large islands into seven small islands.

Mexico

Pepper isn't just a spice you shake onto your food. It's also the name of a moth that lives in Britain. The wings of the peppered moth are covered with black and white splotches. It looks like someone sprinkled them with salt and pepper.

Like all species, each individual peppered moth is unique. Some are born with lighter wings, some with darker wings.

Before the Industrial Revolution (in the late 18th century), the light-colored moths blended in well with the light-colored lichen (a fungus with an algae living in it, pronounced LIE-kin) growing on the trees where the moths rested during the day. This camouflage helped the moths avoid being seen and eaten by birds—their number one enemy and the main selective pressure working on the moths. In 1848, 98 percent of the moths were light-colored.

But as the Industrial Revolution continued, coal dust from factories in many parts of Britain killed off a lot of the lichen and blackened the bark underneath. Against this new, darker backdrop, the lighter moths became easier for birds to spot. Black moths (which had existed before in very small numbers) became more and more common, and in just 50 years, 95 percent of the moths found in the wild were black.

Think It Over

1. How did natural selection cause evolution in the peppered moths in Britain?

2. Today people in many industrialized parts of Britain have begun cleaning up their factories. As a result, the lichens are returning and the trees are becoming less black. How do you think this change will affect populations of peppered moths?

Why Is Biodivers

The activity pages in this section illustrate the complex and amazing ways biodiversity supports ecosystems, affects human health and societies, and helps sustain life on Earth. For the corresponding activities, see pages 180-237 in the Educator's Guide.

ty Important?

"*Earth's animals, plants,
fungi, and microbes are
essential to human health and
well-being—and to life itself.*"

—Francesca T. Grifo, botanist

POLLINATION PUZZLE

Why do people talk about "the birds and the bees" when referring to reproduction? It's because birds and bees play such an important role in plant reproduction. When birds, bees, and other creatures visit flowers to feed on nectar, they transfer pollen from the male parts of flowers to the female parts. Some plants can be pollinated by many different animals. But other plants have evolved so that only one pollinator can reach the pollen in their flowers. Using the clues at the bottom of the page, see if you can match the plants below with the animal that's best suited to pollinate their flowers.

Plants

Strelitzia (Bird-of-Paradise)

Trumpet creeper

Saguaro cactus

Angraecum orchid

Pollinators

Mexican long-nosed bat

Ruby-throated hummingbird

Hawk moth

Ruffed lemur

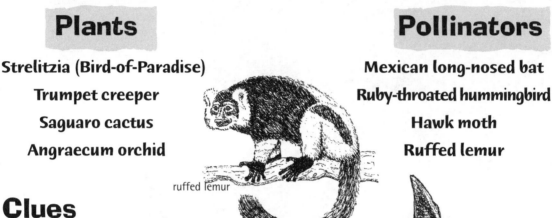

ruffed lemur

ruby-throated hummingbird

Mexican long-nosed bat

hawk moth

Clues

- Ruby-throated hummingbirds are active during the day.

- Strelitzia, also known as bird-of-paradise, has groups of flowers that can be more than one foot long.

- Strelitzia flowers are multicolored, trumpet creeper flowers are bright red, and saguaro and Angraecum orchid flowers are white.

- Ruby-throated hummingbirds have small bodies and long beaks.

- Hawk moths have a very long tongue that they keep coiled beneath their head until they need it.

- Hummingbirds are attracted to red flowers.

- White flowers are easier to see at night.

- Ruffed lemurs are about two feet long.

- Bats and hawk moths are nocturnal. Lemurs are active during the day.

- The nectar in Angraecum orchid flowers is at the end of a tube that can be up to 12 inches long.

- Trumpet creeper flowers are shaped like little trumpets.

- Mexican long-nosed bats have a 3-inch-long tongue.

Soon after a plant is pollinated, it produces seeds. Then what? Ideally, those seeds get scattered so they grow in many new places and don't compete with the parent plant for space, sunlight, nutrients, or water. Plants have developed all sorts of ways to disperse seeds. For example, the burdock plant produces seeds called burrs that stick to animals' fur or our clothing when we brush against the seeds. That way the burdock seeds get a free ride to new possible growing places. There are three key ways that seeds are dispersed: by water, wind, and animals. Do you think that seeds carried by water look like seeds carried by the wind? How do you think the two seeds would differ? Try to name some of the characteristics of seeds dispersed by each of the following means.

Water

Would a seed that's dispersed by water have to float or sink?
Describe features it might have.

....

....

....

Wind

What would a seed that's dispersed by wind look like?
Describe features it might have.

....

....

....

Animals

How do animals disperse seeds? What do you think a seed that's
dispersed by animals would look like? Describe features it might have.

....

....

....

Now see if you can match the following seeds with one of the three means of dispersal described on the previous page. How are these seeds spread? Write *wind*, *water*, or *animal* under each of the seed descriptions below.

a. red maple tree seed (small with wings)

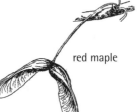

red maple

b. coconut palm seed (large and hollow)

c. pin oak tree acorn (small and nutritious)

pin oak

d. red mangrove tree seed (long and lightweight)

e. wild cherry tree seed (small, contained within fruit)

f. dandelion seed (small, connected to dandelion tufts)

dandelion

The organisms that live in any ecosystem are connected in complex food webs. Look at the sample food web diagram below, then create your own food web for the North American prairie using the information provided. You can create your web with just words or with a combination of pictures and words.

Start by writing the words "prairie rattlesnake" (or draw a picture of one). Then find something that the rattlesnake eats, such as a prairie dog. Write "prairie dog" (or draw a picture of one), and then draw an arrow from the prairie dog to the rattlesnake. The arrows should always point in the direction in which the energy is flowing. For example, the arrow should point to the animal who eats another plant or animal. Keep adding and connecting the clues until you've used them all. Your final drawing should look something like a spider's web.

When you've finished, look at your diagram. Can you tell which animals are the top predators? Would you say rattlesnakes depend on the sun for food? Why or why not?

prairie rattlesnake

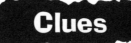

Clues

- **Prairie rattlesnakes eat pocket gophers, mice, and prairie dogs**.

- **Pocket gophers eat grass seeds and roots**.

- **Harvest mice eat grasses and grass seeds**.

- **Prairie dogs eat grasses**.

- **Badgers eat prairie dogs, gophers, and mice**.

- **Coyotes eat mice, pocket gophers, prairie grasses, and fruits**.

- **Red-tailed hawks eat snakes, prairie dogs, pocket gophers, and mice**.

- **Decomposers (such as worms, fungi, beetles, and bacteria) break down all plants and animals when they die**.

- **Decomposers release nutrients into the soil that are then used by plants**.

- **Grasses use sunlight to make their own food through a process called photosynthesis**.

Sample Aquatic Food Web

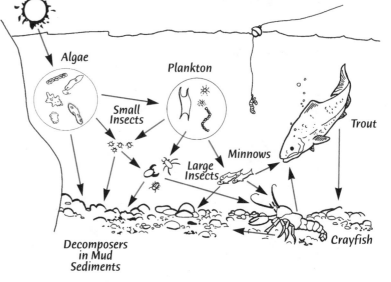

This drawing has been adapted from Olga N. Clymire. 1994. *A Child's Place in the Environment*. Sacramento, Calif.: California Department of Education.

66

The Congo region of Central Africa teems with diversity. Coastal mangrove swamps give way farther inland to rain forests, tropical dry forests, and grasslands. Many mammals, including African elephants and duikers (small antelopes), roam the grasslands. Giant termite mounds rise above the horizon. Columns of driver ants, which can strip a carcass of all its meat in minutes, look for food. If that already sounds like a lot of diversity, think of all the differences that exist within a single species! Each individual elephant, for example, differs in height, tail length, and trunk size.

The above paragraph describes the three major levels of biodiversity—ecosystems, species, and genes. In the chart below, list every example you can find from the paragraph above that matches one of the levels of biodiversity. When you've finished, try to list examples of each of these levels of biodiversity in your own state.

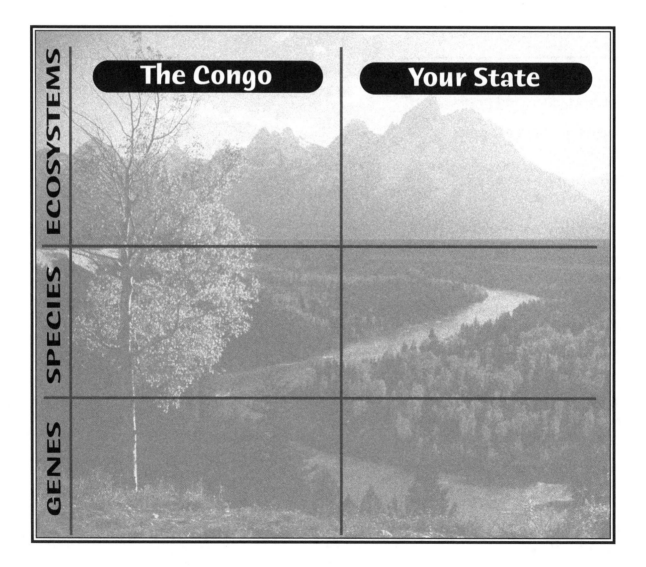

ECOSYSTEMS | SPECIES | GENES

The Congo   **Your State**

North American deserts stretch from the Rio Grande region of Texas and northern Mexico to southern California, and northward through the lowlands of the Great Basin to Washington, Idaho, and Montana (see the map below). The plants and animals that inhabit these regions have many unique adaptations that help them survive the dry and often hot conditions. Reorder the following sets of sentences so they form logical paragraphs about a desert plant or animal.

a.

_____Thanks to this adaptation, when the sun is strong, cacti are able to retain more of their water than plants with leaves.

_____One reason is that they have spines instead of leaves.

_____Of all the plants in the desert, cacti are probably the most successful.

_____Spines don't lose as much water as leaves do.

b.

_____When the rats digest this food, they're able to combine the food's hydrogen with oxygen to form water (H_2O).

_____Kangaroo rats have a unique way of obtaining water in the desert.

_____While their food (mostly seeds) is very dry, it contains hydrogen in varying amounts.

_____By creating water in this way and by having lots of special ways of conserving water, kangaroo rats can live in the desert without drinking any water at all!

c.

_____These lizards have special scales that form fringes on their toes.

_____As a result, they're able to travel beneath the surface of the sand.

_____The fringes help the lizards push through loose sand.

_____Ocellated sand lizards are remarkably well-suited for living in sandy habitats.

1 = Great Basin Desert

2 = Mojave Desert

3 = Sonoran Desert

4 = Chihuahuan Desert

BELIEVE IT OR NOT!

PUTTING THE A·N·T IN PLANT

Some plants, such as acacia trees, make special chambers that ants live in. The ants protect the plant by warding off other insects and plants. And the ants' wastes are used by the plants as a source of nutrients.

KELP HELP

Kelp forms unique underwater forests, but it's sometimes eaten by sea urchins and other plant eaters. Lucky for the kelp, sea otters eat these herbivores and thereby help keep the kelp forests growing strong.

TURN UP THE HEAT

Skunk cabbage heats up a few degrees as it grows to maturity. That makes its flowers more fragrant and attracts the bees and flies that pollinate it.

skunk cabbage

LIFE DOWN UNDER

Scientists are finding that life thrives in places where no one thought it could. Bacteria have been found near hot volcanic vents in the ocean depths where the temperature is over 600°F. Bacteria also have been found in solid rock that is 9,180 feet beneath the Earth's surface.

BLUE GENES

Sclater's black lemurs are the only primates other than humans that have blue eyes. All other nonhuman primates, such as monkeys, chimpanzees, marmosets, and gorillas, are brown eyed.

BUGGIN'

Venus's flytraps, pitcher plants, and sundews are plants that grow in areas that are low in soil nutrients. And they all survive on the same dietary supplement—insects! Each of these plants has developed a unique way to catch insects and then digest them.

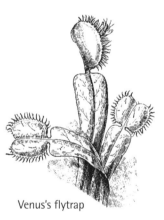

Venus's flytrap

MEAL MATES

Coyotes and badgers sometimes team up to find food. The coyote appears to use its keen sense of smell to find rodents burrowed underground. Then the badger digs them up, and the coyote and badger share the meal.

AMAZING MIGRATIONS

Thousands of monarch butterflies winter in the forests of central Mexico each year. Each spring they migrate north. Some of their offspring eventually reach the northern United States and southern Canada. In the fall the offspring head back to Mexico.

monarch butterfly

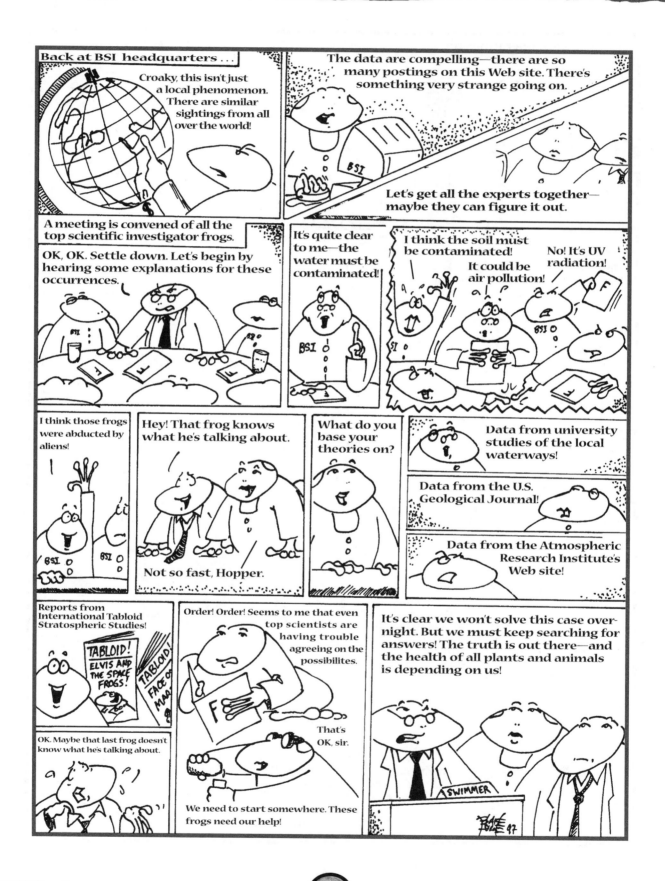

GET ON YOUR CASE!

Now it's your turn to investigate an unsolved mystery. Select one of the cases below and see how well you can answer the questions.

Killer Viruses

The Ebola virus and other deadly viruses have struck several villages in Africa in the last few decades. Could a massive outbreak strike the whole world soon?

- Who's doing research on these deadly viruses?

- What books, newspapers, magazines, and movies have dealt with the topic?

- What's a possible connection between these viruses and rain forest destruction?

- What other reasons have been given for why these viruses are becoming more common and more deadly?

- What's your opinion on this topic? Do you think a deadly outbreak is coming? Or do you think this case is plagued with doubt?

Case #1

Fertility Crisis

Some scientists think the capacity of wildlife and people to reproduce may be reduced by chemicals in the environment. The culprit, says scientist Theo Colborn, may be chemicals that mimic female hormones, making some males infertile, and causing other health problems.

- What are the sources of the chemicals Theo Colborn blames for increased sterility?

- Which species are known to have been affected by these chemicals?

- Does anybody disagree with Dr. Colborn's ideas? Who? What do they say about the topic?

- What kind of research is happening now on this topic?

- What's your opinion on this topic? Are male members of certain species in danger of becoming sterile? Or do you think this case is just mimicking a crisis?

Case #2

Global Warming

Many scientists now agree that the world is experiencing a period of global warming. And some say that it might be causing lots of environmental and health problems—such as increasing the spread of diseases carried by mosquitoes or rodents and causing some species to become extinct.

- List some of the human health problems linked to global warming. What are your sources of this information?

- How might the health of other animals and plants be affected by global warming? Who says?

- Who has a different opinion on these issues? What do they believe?

- What are the credentials of the people on different sides of the debate?

- What's your opinion on this topic? Do you think global warming is beginning to cause global health problems? Or do you think this case is full of hot air?

Case #3

Canaries in the Coal Mine

Some people say that animals that are being hurt by environmental changes are like "canaries in the coal mine." As indicator species, these animals may be providing early warnings about what will eventually happen to other species, including humans, if we don't change our ways.

- Where did the expression "canaries in the coal mine" come from?

- Name some of the species that are considered "canaries in the coal mine," or indicator species. What environmental changes are their health problems associated with?

- Does anybody disagree with this portrayal? Who?

- What's your opinion on this topic? Do you think threatened animals are indicators of our future? Or do you think this case is for the birds?

Case #4

Frogs Revisited

You read about the deformed frogs discovered in Minnesota and other parts of the world. Aren't you curious about the current status of this intriguing case?

- How many different groups can you find that are researching frog health? Which countries are participating?

- What kinds of information are people now getting about the frogs? What problems with frogs have researchers uncovered?

- List as many different explanations for the frog deformities as you can find. What are the sources of your information?

- What effects might frog deformities have on other species?

- Might humans be affected by the same environmental changes?

- Have any groups said they don't believe there's a frog crisis? Which ones?

- What's your opinion on this topic? Do you think frogs are suffering from environmental contamination, or does another explanation leap out at you?

Case #5

Sheltered from the Sun

Have you ever felt sad, or blue, during the cold, dark days of winter? Scientists think these "winter blues" may be caused when people are not exposed to enough sunlight. In the northern parts of the United States, many people are trying light therapy to help them overcome this SAD (Seasonal Affective Disorder) feeling.

- Why does sunlight play an important role in our feeling of well-being?

- What sort of therapy is being tested to help people with SAD?

- Do most scientists believe people are really affected by seasonal changes?

- What's your opinion on this topic? Do you think the sun really influences our health and well-being? Or do you think Seasonal Affective Disorder is a sad excuse for the truth?

Case #6

Cut out the cards below and give one card to each group.

INSECTS, BIRDS, AND BATS HELP POLLINATE.

In their daily search for food, bees and other insects as well as some birds and bats, end up moving pollen from plant to plant. While stopping at a flower for a sip of sweet nectar, the animals get dusted with pollen. When they fly to another flower, some of that pollen brushes off and the pollinated flowers are then able to make seeds. Pollination not only helps wild plants but is also important for crop plants. Most of our crops depend on these natural pollinators.

SOME SPECIES HELP CONTROL POTENTIAL PESTS.

Predators often help keep populations of potential pests in check. For example, birds, bats, and dragonflies are responsible for eating millions of insects that might otherwise gobble up crops or give us itchy bites.

SOME ORGANISMS DECOMPOSE ORGANIC MATTER.

Some living things, called decomposers, get the food they need by feeding on dead things. Decomposers not only keep dead organisms from piling up, they also make the nutrients in the dead organisms available to living plants and animals. Any nutrients they use to build their own bodies become available to other animals that eat them. Also, the nutrients that pass through the decomposers as waste end up in the soil in simpler forms that plants can absorb with their roots. Imagine what life would be like without decomposers!

WETLANDS HELP CLEAN WATER.

If you poured dirty water through a filter, you would expect cleaner water to come out. A similar thing happens in nature when water passes through a wetland. By slowing the flow of dirty water as it goes by, the vegetation growing in a wetland traps some of the pollutants and sediments. But plants aren't the only living things that clean water. Aquatic animals, such as oysters, that pump water through their bodies to filter out food for themselves also end up cleaning the water they live in.

Wetlands are areas that have waterlogged soils or are covered with shallow water either all the time or off and on. Freshwater and salt marshes and swamps, as well as bogs, are all wetlands.

freshwater marsh   **bog**   **salt marsh**   **mangrove swamp**

PLANTS HELP CONTROL EROSION AND FLOODING.

Have you ever seen rainwater rushing down a hillside that has little plant cover? With little vegetation (plant cover) to slow it down and absorb it, water sweeps away soil at a rapid rate. Plants slow down water, allowing the soil to soak it up. So plants help prevent both erosion and flooding.

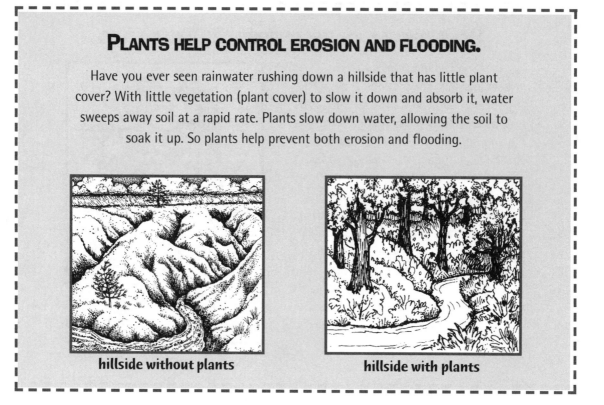

hillside without plants   **hillside with plants**

PLANTS CONVERT THE SUN'S ENERGY INTO ENERGY WE CAN USE.

Although the first warm days of spring may make you feel energetic, we humans (and other animals) can't get the energy we need to fuel our activities directly from the sun. Instead we rely directly or indirectly on plants for energy. Green plants capture the sun's energy and convert it to starch and sugar through a process called photosynthesis. They store some of the energy in their leaves and stems. When animals eat plants, the animals get the energy that the plants stored and use it or store it. And when animals eat animals that ate plants, they then get the energy passed along. Without green plants we'd all go hungry!

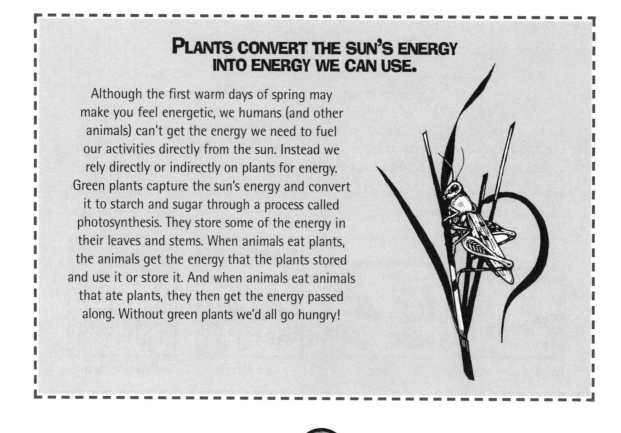

PLANTS AND ANIMALS WORK TOGETHER TO HELP MAINTAIN THE BALANCE OF GASES IN THE AIR.

Plants and animals continuously cycle gases among themselves, the soil, and the air. For example, plants take in carbon dioxide from the air and then release oxygen into the air during photosynthesis. Animals, including humans, use oxygen in respiration and release carbon dioxide into the air. The carbon cycle is even more complicated because plants also respire, using oxygen and releasing carbon dioxide. Water vapor and other gases, such as nitrogen, also cycle from the atmosphere to animals and plants, to the soil, and back again. Without living things, the air just wouldn't be the same!

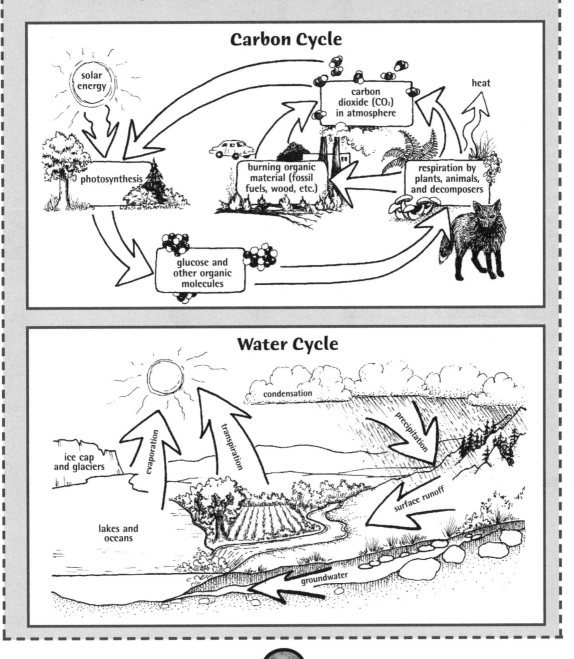

Carbon Cycle

solar energy

carbon dioxide (CO_2) in atmosphere

heat

photosynthesis

burning organic material (fossil fuels, wood, etc.)

respiration by plants, animals, and decomposers

glucose and other organic molecules

Water Cycle

condensation

precipitation

ice cap and glaciers

evaporation

transpiration

surface runoff

lakes and oceans

groundwater

78

THE NATURE OF POETRY

In Time of Silver Rain
by Langston Hughes

In time of silver rain
The Earth
Puts forth new life again,
Green grasses grow
And flowers lift their heads,
And over all the plain
The wonder spreads
    Of life,
    Of life,
    Of life!
In time of silver rain
The butterflies
Lift silken wings
To catch a rainbow cry,
And trees put forth
New leaves to sing
In joy beneath the sky
As down the roadway
Passing boys and girls
Go singing, too,
In time of silver rain
    When spring
    And life
    Are new.

—From *Collected Poems* by
Langston Hughes. Copyright © 1994
by the Estate of Langston Hughes.
Reprinted by permission of
Alfred A. Knopf, Inc.

The Panther
by Ogden Nash

The panther is like a leopard
Except it hasn't been peppered.
Should you behold a panther
crouch,
Prepare to say Ouch.
Better yet, if called by a panther
Don't anther.

—From *Verses from 1929 on*
by Ogden Nash. Copyright © 1940
by Ogden Nash. First appeared in
The Saturday Evening Post.
Reprinted by permission of
Little, Brown and Company.

Rain Forest
by Joseph Richey

tall
lush
rain
forest
dripping in the morning
wild orchids banana flowers
thick vines drape los palos del sol & great white cedar;
others w/ five foot green elephant ears flopping,
hundreds of butterflies, orange caterpillars, blue
      mosquitoes, pink mushrooms,
& industrious leaf cutting ants commuting w/ hunks of
      green to store in infinite
ant caverns & nourish among the fungus formed
      therein;
tiny crimson roots twist around larger roots twisting
      around thicker branches spiraling around
larger trunks of trees, disneyesque the organic
      biospheric plumbing in the world
& a black butterfly w/ red striped wings flutters
      without a sound
through the billions of green leaves quivering moist in
      the patchy sunlight

—From *Earth Prayers: From Around the World,
365 Prayers, Poems, and Invocations for Honoring
the Earth* by Elizabeth Roberts and Elias Amidon,
eds. Copyright © 1991, Harper San Francisco.

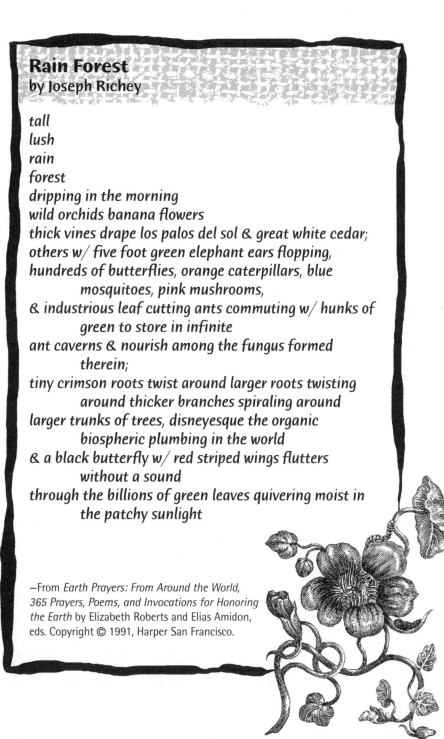

What Makes the Grizzlies Dance
by Sandra Alcosser

June and finally snowpeas
sweeten the Mission Valley.
High behind the numinous meadows
ladybugs swarm, like huge
lacquered fans from Hong Kong,
like the serrated skirts
of blown poppies,
whole mountains turn red.
And in the blue penstemon
the grizzly bears swirl
as they bat snags of color
against their ragged mouths.
Have you never wanted to spin
like that on hairy, leathered
feet, amid the swelling berries
as you tasted a language
of early summer? Shaping
the lazy operatic vowels,
cracking the hard-shelled
consonants like speckled
insects between your teeth,
have you never wanted
to waltz the hills
like a beast?

—From *Imaginary Animals*:
*Poetry & Art for Young
People* by Charles Sullivan,
ed. Copyright © 1996,
Abrams Publishing.
Reprinted by
permission of
the author.

For a Coming Extinction
by H. S. Merwin

Gray whale
Now that we are sending you to The End
That great god
Tell him
That we who follow you invented forgiveness
And forgive nothing

I write as though you could understand
And I could say it
One must always pretend something
Among the dying
When you have left the seas nodding on their stalks
Empty of you
Tell him that we were made
On another day

The bewilderment will diminish like an echo
Winding along your inner mountains
Unheard by us
And find its way out
Leaving behind it the future
Dead
And ours

When you will not see again
The whale calves trying the light
Consider what you will find in the black garden
And its court
The sea cows the Great Auks the gorillas
The irreplaceable hosts ranged countless
And fore-ordaining as stars
Our sacrifices
Join your word to theirs
Tell him
That it is we who are important

—From *The Lice* by H. S. Merwin. Copyright
© 1963, Georges Borchardt, Inc. Reprinted
by permission of Georges Borchardt, Inc.
for the author.

A Song for New-Ark
by Nikki Giovanni

When I write I like to write...in total silence...Maybe total...silence...is not quite accurate...I like to listen to the notes breezing by my head...the grunting of the rainbow...as she bends...on her journey from Saturn...to harvest the melody...

There is no laughter...in the city...no joy...in the sheer delight...of living...City sounds...are the cracking of ice in glasses...or hearts in despair...The burglar alarms...or boredom...warning of illicit entry...The fire bells proclaiming...yet another home...or job...or dream...has deserted the will...to continue...The cries...of all the lonely people...for a drum...a tom-tom...some cymbal...some/body...to sing for...

I never saw old/jersey...or old/ark...Old/ark was a forest...felled for concrete...and asphalt...and bridges to Manhattan...Earth acres that once held families...of deer...fox...chipmunks...hawks...forest creatures...and their predators...now corral business...men and women...artists...and intellectuals...People...and their predators...under a banner of neon...graying the honest Black...cradling the stars above...and the Earth below...turning to dust...white shirts...lace curtains at the front window...automobiles lovingly polished...Dreams...encountering racist resistance...New-Ark knows too much pain...sees too many people who aren't special...watches the buses daily...the churches on Sunday...the bars after midnight...disgorge the unyoung...unable...unqualified...unto the unaccepting...streets...I lived...one summer...in New-Ark...New-Jersey...on Belleville Avenue...Every evening...when the rats left the river...to visit the central ward...Anthony Imperiali...and his boys...would chunk bullets...at the fleeing mammals...refusing to recognize...the obvious...family...ties...I napped...to the rat-tat-tat...rat-tat-tat...wondering why...we have yet to learn...rat-tat-tats...don't even impress...rats...

When I write I want to write...in rhythm...regularizing the moontides...to the heart/beats...of the twinkling stars...sending an S.O.S....to day trippers...urging them to turn back...toward the Darkness...to ride the night winds...to tomorrow...I wish I understood...bird...Birds in the city talk...a city language...They always seem...unlike humans...to have something...useful...to say...Other birds...like Black Americans...a century or so ago...answer back... with song...I wish I could be a melody...like a damp...gray... feline fog...staccatoing...stealthily...over the city...

Free verse

Any number of open lines with no set rhyme or pattern.

Sun sets gently
through the horizon cushion
to be absorbed
beyond view.

Cinquain

Verses with the following pattern:

Line 1—one word title
Line 2—two words describing title
Line 3—three words showing action
Line 4—four words showing a feeling about the title
Line 5—one word (simile or metaphor for the title)

Swallows
sleek, deft
diving, soaring, flying
bringing joy to Earth
dancers.

Haiku

A type of poetry from Japan with a very structured pattern.

Line 1— 5 syllables   trees bend with strong wind

Line 2— 7 syllables   gusts and torrents blow so hard

Line 3— 5 syllables   yet ever rooted

Diamante

A poem written in the shape of a diamond in a set use of words. Often the first half of the poem is the opposite of the second half.

noun
adjective adjective
participle participle participle
noun noun noun noun
participle participle participle
adjective adjective
noun

turtle
bulk cumbersome
crawling creeping dragging
shell legs paws fur
running racing leaping
quick slick
cheetah

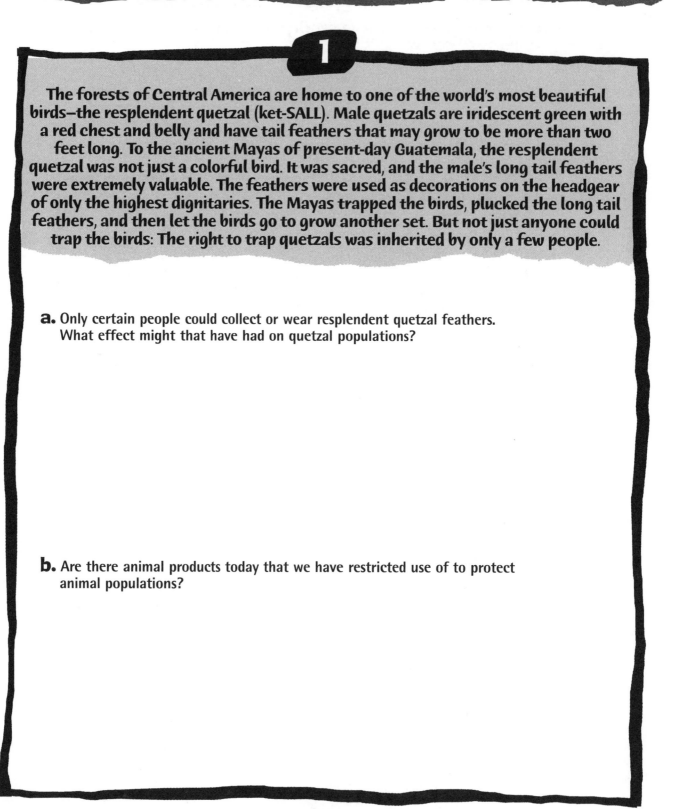

1

The forests of Central America are home to one of the world's most beautiful birds—the resplendent quetzal (ket-SALL). Male quetzals are iridescent green with a red chest and belly and have tail feathers that may grow to be more than two feet long. To the ancient Mayas of present-day Guatemala, the resplendent quetzal was not just a colorful bird. It was sacred, and the male's long tail feathers were extremely valuable. The feathers were used as decorations on the headgear of only the highest dignitaries. The Mayas trapped the birds, plucked the long tail feathers, and then let the birds go to grow another set. But not just anyone could trap the birds: The right to trap quetzals was inherited by only a few people.

a. Only certain people could collect or wear resplendent quetzal feathers. What effect might that have had on quetzal populations?

b. Are there animal products today that we have restricted use of to protect animal populations?

2

Early settlers to the American West sometimes wore buckskin or elkskin pants to protect themselves against the cold. The pants were also an excellent way to keep thorny shrubs from scratching their legs. In addition, the settlers wore knee-high leather boots over woolen socks—a combination that kept them warm and protected their legs from rattlesnakes!

a. Why did the settlers choose buckskin and elkskin for their pants?

b. Where do your clothes come from? Do they have anything to do with nature?

3

In ancient Rome, senators, judges, and the wealthy were the only people who wore purple. That's because the only known source of purple dye was murex snails. These marine snails live in rocky, shallow water, and each one produces only a tiny amount of dye. Huge piles of discarded shells in certain areas of the Mediterranean region testify to the tens of thousands of murex snails that were killed for their purple dye.

a. Why was purple reserved for high-ranking and wealthy people in ancient Rome? Why can all people wear purple today?

b. Are there things that only high-ranking and wealthy people wear today? What makes these things so valued?

4

With their bright red coats and tall hats, the guards of Buckingham Palace in England may be the most photographed soldiers in the world. Their distinctive hats are covered in Canadian bear fur in a tradition that is hundreds of years old. In the 17th century, elite French soldiers called grenadiers began attaching bits of bear fur to their hats. Over time, the soldiers added more and more fur to their hats until, by the 18th century, almost the entire hat was covered by bear fur. When British soldiers defeated Napoleon's grenadiers at the Battle of Waterloo in 1815, they earned the right to wear the French soldiers' bear fur. Today the Buckingham Palace guards and other British guards still wear bear fur hats, although the British are hoping to find a synthetic substitute for the fur.

a. Why do you think grenadiers first wore hats covered in bear fur?

b. Do people in your community use animals as symbols of strength? If so, in what ways?

5

In many parts of Africa, both lions and leopards have been used in art to represent a variety of characteristics. For example, some groups in Mali carved wooden masks into the image of a lion. The masks were worn in special ceremonies in which dancers would imitate lions. The dance presented a model for how human leaders should behave.

a. What characteristics of lions do you think Mali mask-makers respected and wanted to imitate?

b. How are animals or plants used as symbols in art or entertainment in your life? What characteristics of these animals do people prize? Give examples.

6

Few foods are as rare or as special in Italy as the wild truffle. This mushroom-like food, which grows mostly on the roots of oak trees, adds an earthy, pungent taste to pasta, meats, and other dishes. Truffles are tasty, but they're not easy to find. That's because truffles grow in unpredictable places—not to mention underground! Truffle hunters have to be experts at looking for truffle clues. Traditionally, many of them have used dogs or even pigs to help sniff out the treats.

a. If a truffle hunter did not have the help of a dog or pig, what skills would he or she need to find truffles? List as many as you can.

b. Have you ever collected food from the wild? If so, what kinds? Where do you get most of your food? If it comes from a store, where does the store get it?

Ofelia (age 12)

"I used to think picking amaranth was fun. But these days I get worried. What if my friends from school see me picking amaranth? What will they think?"

Friends of Ofelia

"Ofelia has a cool family. We just wish she'd tell us more about her culture and her traditions."

Mother

"I think it's good for Danny and Ofelia to come out and pick amaranth greens every year. They need to learn about our family's traditions. They need to know their culture."

Friends of Danny

"Danny's family is kind of strange. We don't understand why they do stuff like collect plants near the highway. Danny should just hang out with us."

Grandmother

"We should always harvest some foods like I used to do when I was young. Danny and Ofelia should grow up with respect for the old ways. They should know where some of their traditional food comes from."

Highway Patrol Officer

"We saw these people picking plants on the side of the road and told them they should stop. It isn't safe. And besides, some of the plants might be sprayed with pesticides."

Danny (age 14)

"If you ask me, I say amaranth is a weed, not a food. I'd rather be out with friends than picking that stuff. Besides, we should get our food from the grocery store like other people do."

Mr. Williams

"I teach health at Ofelia's middle school. The kids here eat too much junk food. I think they should eat more fresh fruits and vegetables. And I think it's great that families collect wild foods."

Father

"I'm not Native American, but I respect the culture of my wife and her family. Still, if the kids don't want to go out and pick amaranth, I don't think they should have to. They should be who they want to be."

WHAT IS AMARANTH?

Amaranth is the general name given to over 50 species of plants in the genus Amaranthus. Used for thousands of years by cultures throughout the world, the different species of amaranth provide food, decoration, and religious articles in ceremonies.

Amaranth was a main food source for many ancient civilizations, including the Indian tribes of North America, and the Aztec, Maya, and Inca civilizations of Central and South America.

Amaranth grain (seed) is full of nutrients. It is higher in protein, fiber, calcium, and iron than most grains. That's probably why it was so important to so many different cultures and why you can find things made with amaranth grain in stores today. The grain is used in breakfast cereals, pastas, breads, cookies, and many more foods. When heated, amaranth seeds burst like popcorn, making them a tasty, nutritious snack.

The leaves are also very high in nutrients. When they are young, the leaves can be boiled and eaten as greens. Amaranth leaves can also be used to feed animals.

Not all amaranth is grown for food. Because it has such colorful flowers, some people plant amaranth in their gardens and still others grow it for use in religious ceremonies. In some parts of the country, amaranth grows along roads and fences and in fields and gardens. It is often called "pigweed," and even these varieties can be eaten by people. Because of its incredible versatility and highly nutritious grain and leaves, amaranth is grown in Europe, India, China, Southeast Asia, and North and South America.

amaranth

Classified Information
Do not share this information with other teams!

Soil is a mixture of mineral particles, air, water, microorganisms and other organic matter (material derived from living things). The materials that make up soil form layers. Hundreds of years may be required to form just a few inches of soil. Soil helps to purify water by filtering out some of the suspended solids (floating "dirt" particles) as they flow through the different soil layers. The makeup of the soil determines how well it will act as a filter. Soil also helps to remove chemical contaminants such as fertilizers and pesticides. Many minerals in the soil can chemically bond with contaminants, which are then stored in the soil and prevented from flowing into nearby waterways. As a result of chemical reactions, the soil can also help "detoxify" certain chemicals, making them less harmful to living things.

Materials

clear funnel or clear plastic soda bottle with the bottom cut off and the label removed, clear plastic cup, tall jar or flask, cotton balls or toilet paper, activated charcoal, sand, potting soil, water

What to do

1. Pack the funnel approximately one-third full with cotton balls.

2. Place a layer of charcoal on top of the cotton balls. Then place a layer of sand on top of the charcoal.

3. Place the funnel into the jar or flask. The mouth of the jar should be small enough to keep the funnel off the bottom of the jar. (See diagram on page 94.)

4. Mix one-fourth cup of potting soil with one-half cup of water in the plastic cup. Then slowly pour the water into the funnel.

What happened?

Describe the appearance of the water after filtering and any changes that you can see. Look at the different layers in your funnel. Where did most of the large soil particles get trapped? Where did the fine particles get trapped? (You should notice that the larger particles remained at the top layer while the finer particles were trapped by the activated charcoal.) What do you observe about your samples?

Think about it

How do you explain the results of the experiment? Why do you think some materials are more effective filters than others? How would you describe the "services" that soils provide?

Adapted from *Water Watchers*; used by permission of the Massachusetts Water Resources Authority.

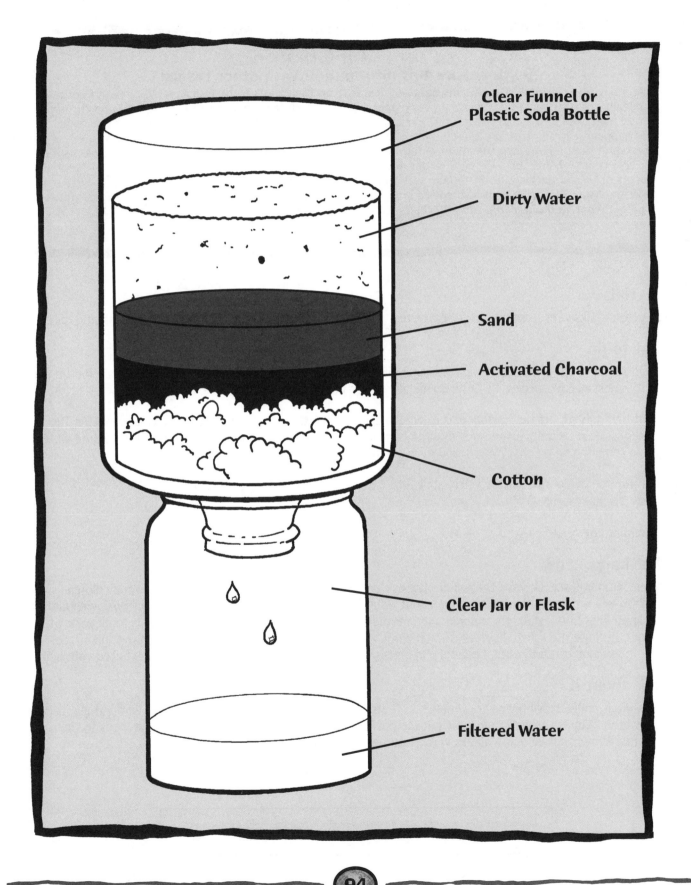

Clear Funnel or Plastic Soda Bottle

Dirty Water

Sand

Activated Charcoal

Cotton

Clear Jar or Flask

Filtered Water

Classified Information
Do not share this information with other teams!

Plants have fine "tubes" inside them that carry water from their roots to their leaves. When water contains toxic pollutants (such as pesticides or heavy metals) those pollutants may also be carried up and through the plant. Many wetland plants store toxic materials in their tissue. This doesn't mean that the toxins disappear—usually they are excreted later. But they are released slowly, in small amounts that are less damaging than a large dose of toxins entering a river, lake, or pond at once. When the wetland plants die, the toxins are released back into the water and soil of the wetland where they may be "captured" by other plants or by soil particles. Even though wetland plants can help absorb and alter some toxins, they aren't able to absorb all toxins. Just as there's a limit to how much water a sponge can absorb, there's a limit to what wetland plants can absorb—especially if toxins enter the wetland in large amounts.

Materials

fresh celery stalks with leaves, a jar or beaker, red or blue food coloring, water, paring knife, magnifying glass

What to do

1. Add several drops of food coloring to a water-filled beaker or jar. The food coloring represents pollution from a toxic substance (pesticides, oil, or heavy metals such as mercury, for example).

2. Cut half an inch off the bottom of a celery stalk, and place the stalk overnight in the colored water. The celery stalk represents plants such as cattails, sedges, and grasses that grow in wetlands. The colored water represents the water that flows through the wetland.

3. On the following day, cut the celery stalk into one-inch pieces so that each team member has a piece.

4. Examine the celery closely.

What happened?

Describe what you see. Observe the tubules (tubes that transport the water). Where do you see the colored water? Do you notice anything interesting about the celery leaves? (As you cut through the celery, you should see colored lines in the stalk. The colored lines are the xylem that transports water and minerals to all parts of the plant. Because the xylem distributes water throughout the plant, you should see color at the edge of the leaves. If you look carefully with a magnifying glass, you should also see the veins in the leaves tinted with color.)

Think about it

Communities are increasingly using wetlands as "natural" water treatment facilities. How do wetland plants help purify water? Why is the water remaining in the beaker still "polluted"? What do you think happens to the pollutants? Why can't we dump all our waste into wetlands?

Adapted from *Discover Wetlands* with permission of the Washington State Department of Ecology, Publications Office, Publication Number 88-16, P.O. Box 7600, Olympia, WA 98504.

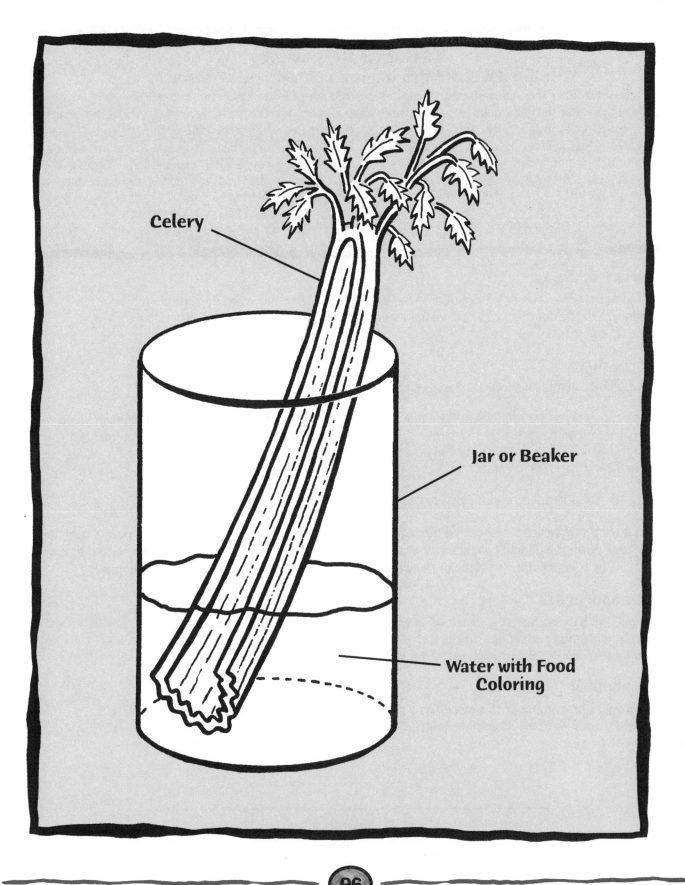

Celery

Jar or Beaker

Water with Food Coloring

> ## Classified Information
> ### Do not share this information with other teams!
> As water flows through wetlands, the grasses slow the speed of the water by simply being in the way. When the water slows, particles of soil and other solids are deposited in the grass, making the water clearer. Larger particles usually settle out first and the smallest particles usually travel the farthest. Wetlands help protect streams, lakes, bays, and other downstream water bodies from a heavy build-up of sediment. They also help protect many aquatic plants and animals. Muddy water covers filter feeders such as clams and oysters, clogs fish gills, smothers fish eggs, "blinds" aquatic animals that hunt for food by sight, and blocks sunlight that aquatic plants and coral animals need to grow.

Materials

several sponges, a doormat or a piece of artificial turf, two flat sheets of wood or plastic similar in size to the doormat, two shallow aluminum trays, soil, two containers of water, props to tilt the models

What to do

1. Set up both boards (or sheets of plastic) on a slant. They need to be at the same angle.

2. Place the doormat (or artificial turf) on one of the boards. Then set the trays at the base of each board. (See diagram on page 98.) These boards represent wetlands. The board with the doormat represents a healthy wetland filled with plants. The other board represents an unhealthy wetland where the plants have died or have been removed.

3. Fill both water containers with equal amounts of water and soil, and mix.

4. Get a team member to stand behind the high end of each board. Now have each of them pour a container of water down the board at the same time and at the same rate. This flow represents water entering the wetland as a stream, flowing through the wetland, and eventually emptying into a lake (the tray).

What happened?

Which wetland had the fastest water flow? In which wetland did more soil settle out? (The model with the doormat or artificial turf should have slowed the water down and trapped more of the larger particles, keeping them from settling out into the tray.)

Think about it

Based on your observations of this model, how do healthy wetlands help provide cleaner water? How could muddy water be harmful to wildlife?

Adapted from *Discover Wetlands* with permission of the Washington State Department of Ecology, Publications Office, Publication Number 88-16, P.O. Box 7600, Olympia, WA 98504.

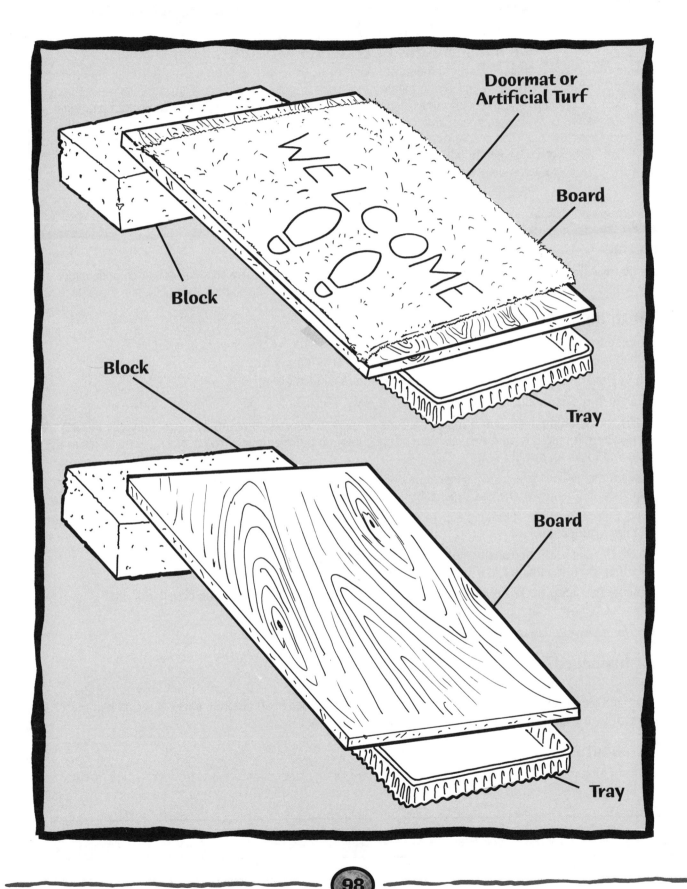

Doormat or
Artificial Turf

Board

Block

Block

Tray

Board

Tray

Classified Information
Do not share this information with other teams!

Water is necessary for life on Earth. Through the water cycle, water moves from the ocean to the atmosphere, to the land, and back to the ocean. Living things also take part in the water cycle. Plants absorb water through their roots and release water into the atmosphere through their leaves in a process called transpiration. Transpiration is the evaporation of water through tiny openings in the leaves. When the water evaporates, any impurities that might be in it stay behind in the plant. In this way water entering the atmosphere is purified. Water released into the atmosphere also contributes to the formation of clouds.

In ecosystems, plants play an important role in determining the amount of water entering the atmosphere, which has a great effect on the climate in an area.

Materials

two large clear plastic cups, six-inch square of wax paper, geranium plant leaf with stem, cobalt chloride paper (available from science supply catalogues–see page 228 in the Educator's Guide), petroleum jelly, paper clip, tape, water

What to do

1. Place a drop of water on a piece of cobalt chloride paper. Observe the change in color. Cobalt chloride paper is used to detect the presence of water.

2. Fill one of the cups with water and apply petroleum jelly to the rim.

3. Straighten the paper clip, and use one end of it to poke a small hole in the center of the square of wax paper.

4. Insert the geranium leaf stem through the hole in the wax paper square. Apply petroleum jelly around the stem where it emerges from the wax paper. Apply enough petroleum jelly to cover any extra space in the hole and make an airtight seal.

5. Position the leaf and wax paper combination directly over the water-filled cup. Gently press down on the wax paper around the rim so the wax paper is held in place by the petroleum jelly. The stem should be in the water.

6. Tape a piece of cobalt chloride paper to the inside bottom of the other cup. Apply petroleum jelly around the rim of the cup.

7. Invert the cup with the cobalt chloride paper over the geranium leaf setup. Gently press the cups together. Do not allow the leaf to touch the cobalt chloride paper.

8. Observe the setup for five minutes. Pay particular attention to the color of the cobalt chloride paper. Leave the setup undisturbed for 24 hours.

9. On the following day make your final observations about the cobalt chloride paper.

What happened?

How do you explain the change in color of the cobalt chloride paper? (As the water made its way through the stem and leaf, it entered into the air of the second jar. Because cobalt chloride paper turns pink in the presence of water vapor, the cobalt chloride paper changed color.)

Think about it

Using the results of the demonstration, what role do you think plants play in the water cycle? How do plants affect local climates? Describe the differences in climate between a forest ecosystem and a desert ecosystem.

Adapted from Activity 4.1 "Plants and the Water Cycle" from Addison-Wesley *Environmental Science: Ecology and Human Impact, 2nd Edition*, by Leonard Bernstein, Alan Winkler, and Linda Zierdt-Warshaw, copyright © 1996 by Addison-Wesley Publishing Company. Reprinted with permission.

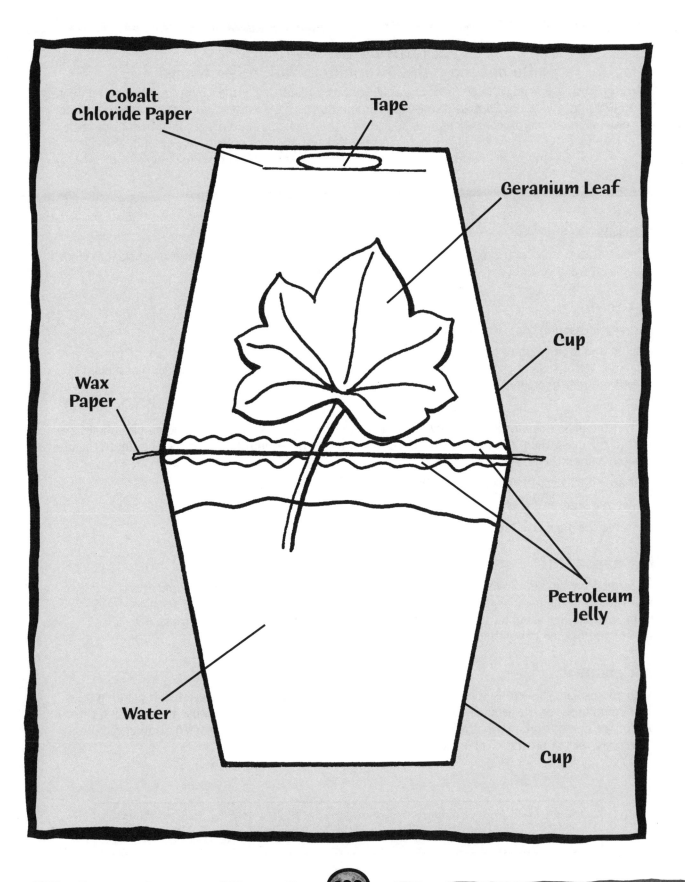

Cobalt Chloride Paper

Tape

Geranium Leaf

Wax Paper

Cup

Petroleum Jelly

Water

Cup

Classified Information
Do not share this information with other teams!

Green plants, like animals, need food. But unlike animals, plants make their own food through a process called photosynthesis. Photosynthesis uses carbon dioxide (CO_2), water, and energy from the sun to produce food and oxygen. Common indoor plants used in homes and offices may help to fight the rising levels of indoor air pollution. NASA scientists are finding plants to be useful in absorbing potentially harmful gases and cleaning the air inside modern buildings.

Materials

large bowl, water, measuring cup, tablespoon, baking soda, drinking glass, lamp, water plant such as elodea or anacharis (available from any pet store that sells fish)

What to do

1. Using a measuring cup, fill a bowl with fresh water.

2. Mix in one tablespoon of baking soda for every two cups of water. (Baking soda is "short" for bicarbonate of soda. It contains carbon and in this experiment it will provide the CO_2 that a plant needs in order to create its own food—to photosynthesize.)

3. Place a water plant, such as elodea, inside a drinking glass. Add enough water to fill up half the glass.

4. Lower the glass sideways into the bowl of water until the glass fills with water and no air bubbles are left in the glass. Then turn the glass upside down in the bowl without letting in air. The top of the glass should rest on the bottom of the bowl.

5. Set up a light near the bowl and aim the light toward one side of the glass.

6. Leave the light on the plant overnight.

7. Observe the plant and the glass of water.

What happened?

What formed the next day? Why? (You should see a bubble the next day. The light stimulates photosynthesis in the plant. As a plant goes through photosynthesis to make food, it releases oxygen. Since oxygen is lighter than water, it rises to the top and is trapped by the glass. After 24 hours, enough oxygen has gathered to form a bubble.)

Think about it

How do you explain the results you see? Why might a city planning board be interested in planting trees in their community, or the people in an office building be interested in having house plants? Using the results of the demonstration, what role do you think plants play in all ecosystems? What other factors are necessary for the process of photosynthesis?

Reprinted with permission, American Forest Foundation, Copyright 1993, 1994, 1995, 1996. *Project Learning Tree Environmental Education Activity Guide Pre K–8.* The complete Activity Guide can be obtained by attending a PLT workshop. For more information, call the National PLT office at (202) 463-2462 or visit their Web site at <www.plt.org>.

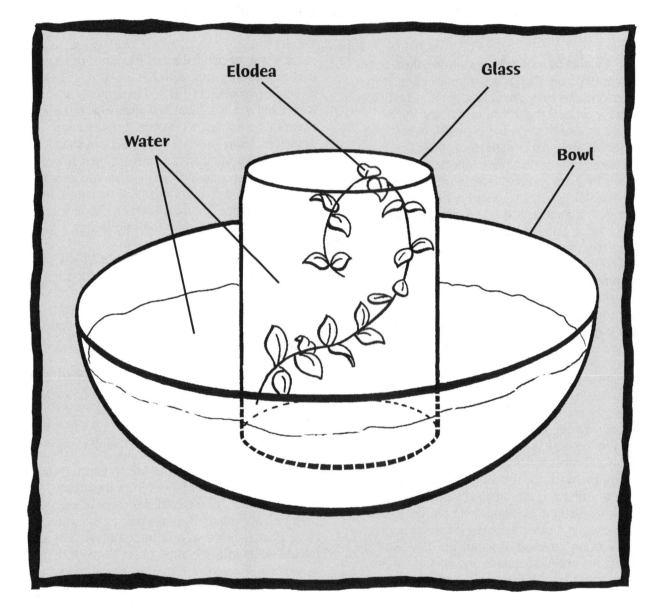

Elodea

Glass

Water

Bowl

BIOFACT

Buried Bacteria Thrive—In the state of Washington, scientists have discovered bacteria living in solid rock more than a half a mile underground. Without sunlight or oxygen, the bacteria seem to thrive on nothing more than rock and water.

102

**Match the number of the station to
the ecosystem service that is described here.**

a. _____Plants have fine "tubes" inside them that carry water from their roots to their leaves. When water contains toxic pollutants (such as pesticides or heavy metals) those pollutants may also be carried up and through the plant. Many wetland plants will store toxic materials in their tissues. This doesn't mean that the toxins disappear— usually they are excreted later. But they are released slowly, in small amounts that are less damaging than a large dose of toxins entering a river, lake, or pond at once. When the wetland plants die, the toxins are released back into the water and soil of the wetland where they may be "captured" by other plants or by soil particles. Even though wetland plants can help absorb and alter some toxins, they aren't able to absorb all toxins. Just as there's a limit to how much water a sponge can absorb, there's a limit to what wetland plants can absorb—especially if toxins enter the wetland in large amounts.

b. _____As water flows through wetlands, the grasses slow the speed of the water by simply being in the way. When the water slows, particles of soil and other solids are deposited in the grass, making the water clearer. Larger particles usually settle out first and the smallest particles usually travel the farthest. Wetlands help protect streams, lakes, bays, and other downstream water bodies from a heavy build-up of sediment. They also help protect many aquatic plants and animals. Muddy water covers filter feeders such as clams and oysters, clogs fish gills, smothers fish eggs, "blinds" aquatic animals that hunt for food by sight, and blocks sunlight that aquatic plants and coral animals need to grow.

c. _____ Soil is a mixture of mineral particles, air, water, microorganisms, and other organic matter (material derived from living things). The materials that make up soil form layers. Hundreds of years may be required to form just a few inches of soil. Soil helps to purify water by filtering out some of the suspended solids (floating "dirt" particles) as they flow through the different soil layers. The makeup of the soil determines how well it will act as a filter. Soil also helps to remove chemical contaminants such as fertilizers and pesticides. Many minerals in the soil can chemically bond with contaminants, which are then stored in the soil and prevented from flowing into nearby waterways. As a result of chemical reactions, the soil can also help "detoxify" certain chemicals, making them less harmful to living things.

d. _____Green plants, like animals, need food. But unlike animals, plants make their own food through a process called photosynthesis. Photosynthesis uses carbon dioxide (CO_2), water, and energy from the sun to produce food and oxygen.

Common indoor plants used in homes and offices may help to fight the rising levels of indoor air pollution. NASA scientists are finding plants to be useful in absorbing potentially harmful gases and cleaning the air inside modern buildings.

e. _____Water is necessary for life on Earth. Through the water cycle, water moves from the oceans, to the atmosphere, to the land, and back to the ocean. Living things also take part in the water cycle. Plants absorb water through their roots and release water into the atmosphere through their leaves in a process called transpiration. Transpiration is the evaporation of water through tiny openings in the leaves. When the water evaporates, any impurities that might be in it stay behind in the plant. In this way water entering the atmosphere is purified. Water released into the atmosphere also contributes to the formation of clouds.

In ecosystems, plants play an important role in determining the amount of water entering the atmosphere, which has a great effect on the climate in an area.

The Irish Potato Famine

In the 1840s, the Irish potato crop was made up almost entirely of one variety known as the "lumper." A fungal infection called "the blight" spread from England and infested the lumpers, wiping out the crop. Within two years, an estimated one million people starved to death and many more were forced to leave the country. Why was this blight so destructive? The Irish people depended almost exclusively on this one variety of potato as a food source. And because the entire crop was genetically uniform and vulnerable to the same disease, there was no way to save it. Many people believe that if a variety of potatoes with different genetic traits had been available, some of them might have survived the blight, preventing much of the hardship.

Discoveries from Nature

A young graduate student on a plant-collecting expedition in Mexico in 1977 was rewarded when some local farmers showed him where a wild species related to corn (also called maize) was thriving in a remote area of the mountains. Because of the species' natural ability to withstand hot, dry weather, its ability to reproduce without the help of a farmer, and its resistance to devastating corn diseases, this plant was an immensely valuable find—particularly to corn farmers in the United States. The genes responsible for these beneficial traits may eventually be bred into U.S. corn crops, saving millions of dollars per year in lost crops. Unfortunately, potentially important wild plants are lost every day as rain forests and other ecosystems are destroyed around the world. The value of these resources cannot be calculated—we don't even know what we are losing.

The Corn Crisis of 1970

In 1970, the genetically uniform corn crops of the United States were hit hard with a disease called the southern corn leaf blight. In one year, the fungus that caused this disease moved through millions of acres of corn from Florida to Minnesota. It eventually destroyed 15 percent of the entire U.S. corn crop, costing U.S. farmers about one billion dollars.

It turned out that a trait that had been bred into corn to enhance the efficiency of its seed production also happened to increase its susceptibility to the deadly fungus. Luckily, the U.S. corn crop was saved the next year because seed companies and plant breeders had saved some older corn varieties in a gene bank and were able to breed new varieties of disease-resistant corn. The process took several years and quite a bit of corn was lost, but using the gene bank worked. However, if we continue to lose crop diversity, we may not be able to find important genes that make plants adaptable to changing environmental conditions.

BIOFACT

Dining on Bumble Bees—The thick-headed fly lays its eggs on bumble bees. When an egg hatches, the larva chews its way into the bee and starts feeding on its insides. This seems to affect the bee's brain, which makes the bee do something it wouldn't normally do—dig into the soil and bury itself. After the bee dies, the fly larva turns into a pupa. It spends the winter safely sheltered underground.

Seed Savers

Gene banks and seed storage laboratories such as the National Seed Storage Laboratory in Fort Collins, Colorado, are facilities that store plant seeds from many different crop varieties. The seeds are kept under very cold conditions in refrigerators or vats filled with liquid nitrogen. Although these banks preserve important genetic material, they are no substitute for preservation of plants in their natural habitats. Living plants are in a constant state of evolution, and seeds, by comparison, are merely snapshots in time. While freezing seeds is an important conservation strategy, it cannot replace conserving plants that adapt over time to new environmental conditions. Fortunately, many people are also collecting seeds that have been passed down as family treasures from one generation to the next. Members of the Seed Savers Exchange, for example, are concerned with protecting and growing seeds that are rarely available in commercial catalogues and are therefore quite vulnerable to extinction. One member of this group grows 400 different varieties of squash in his garden!

Valuable and Endangered U.S. Plants

Although most of the major food crops in the United States come from other parts of the world, there are many plants native to our soils that have agricultural value. Unfortunately, many of these plants are also rare and endangered. According to the Center for Plant Conservation (CPC), a national network of botanical gardens and arboretums working to save America's endangered plants, one-fifth of the 20,000 plants native to the United States are in need of conservation. Included in this figure are the Okeechobee gourd, a very rare relative of cultivated squash and pumpkins that is known to have resistance to certain diseases that commonly attack these crops, and Texas wild rice, which is being studied and used to develop a new strain of edible rice. A recent study by the CPC found that more than 80 percent of the endangered plants in the United States have close relatives that are economically and/or agriculturally important. Based on this study and other research, scientists believe that the related endangered plants could also be useful and important, thus providing another reason to save these plants and study them before we lose them. For more information, visit the CPC's Web site at <www.mobot.org/cpc>.

PLANTING GRID

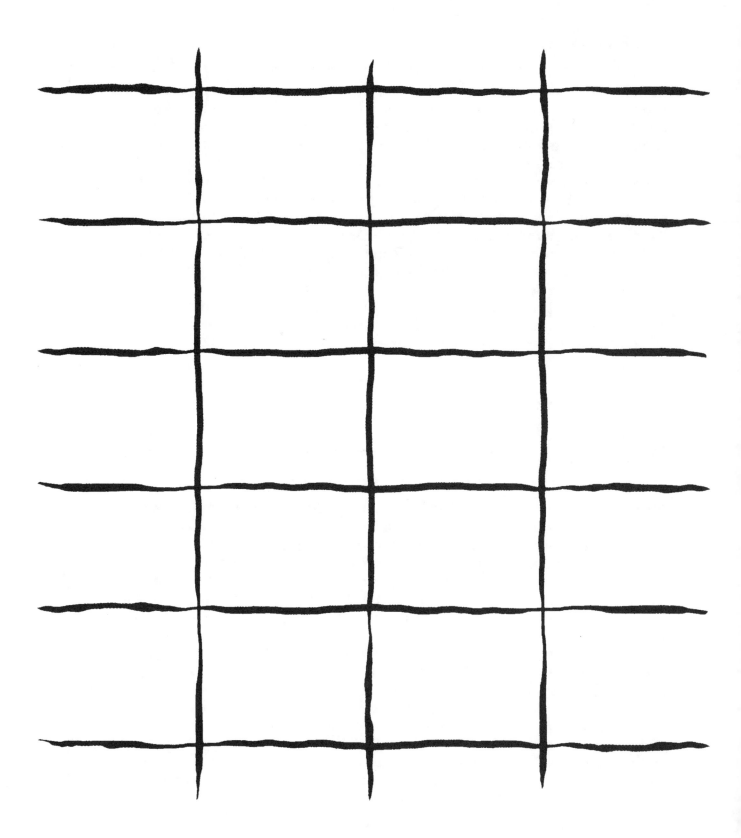

CROP Crisis

CROP Crisis **Hot! Hot! Hot!**
The earth is cracking beneath your feet. It hasn't rained in six weeks! No sprinkler system can compete with such a drought—the worst in over 25 years. Only drought tolerant beans survive.

CROP Crisis **A Moldy Mess**
Where'd all that white stuff come from on your neighbor's bean plants? This bean mold is heading your way, and it can wipe out an entire crop in no time! Only beans resistant to the bean mold survive.

CROP Crisis **Rabbits?**
Who would have thought such cute, furry creatures could be so destructive? They seem to be everywhere, and they've all got their eyes on your leafy beans! Only beans with poisonous leaves survive.

CROP Crisis **Good News!**
It has been a good year—not too hot or too cold, not too dry or too wet, no insect or slug infestations. You get a full harvest.

CROP Crisis **Seeing Spots?**
A devastating spot fungus is attacking your bean plants. Only beans resistant to spot fungus survive.

CROP Crisis **Design Your Own Crisis!**
(Remember to update the "Traits, Yields, and Values" table.)

CROP Crisis **Slug Fest!**
A bean-eating, pesticide-resistant slug from South America has made its way to your fields, bringing thousands of hungry relatives. Only beans resistant to slugs survive.

FOODS EATEN

Day 1

Breakfast	Lunch	Dinner	Snacks

Day 2

Breakfast	Lunch	Dinner	Snacks

Day 3

Breakfast	Lunch	Dinner	Snacks

"Education is the greatest hope
for guaranteeing that our children inherit
a healthy environment. It's their
dreams and commitment that will change
the way future generations think about
the Earth and how we care for it."

**–Kathy McGlauflin,
environmental educator**

What's the Status

The activity pages in this section explore biodiversity in today's world—what's happening to it, why it's threatened, where it's thriving, and how it's being studied. For the corresponding activities, see pages 238-321 in the Educator's Guide.

# of Biodiversity?

"Around the world, biodiversity–defined
as the full variety
of life from genes to species to
ecosystems–is in trouble.
There is not one country,
not one biome, that
remains untouched."

–Edward O. Wilson, biologist

Mammals

Northern Hairy-Nosed Wombat

Red Wolf

Maned Wolf

Giant Panda

Spectacled Bear

Black-Footed Ferret

Cheetah

Indiana Bat

Mexican Long-Nosed Bat

Indri

Aye-Aye

Golden Lion Tamarin

Cotton-Top Tamarin

Uakari (Red and White)

Lion-Tailed Macaque

Wooly Spider Monkey

Proboscis Monkey

Jaguarundi

Ocelot

Margay

Clouded Leopard

Jaguar

Leopard

Tiger

Snow Leopard

Asian Elephant

African Elephant

Hawaiian Monk Seal

Dugong

West Indian Manatee

African Wild Ass

Grevy's Zebra

Prezewalski's Horse

Central American Tapir

Malayan Tapir

Black Rhinoceros

Sumatran Rhinoceros

Javan Rhinoceros

Great Indian Rhinoceros

Babirusa

Pygmy Hippopotamus

Swamp Deer

Gorilla

Mountain Gorilla

Pygmy Chimpanzee

Orangutan

Giant Anteater

Utah Prairie Dog

Delmarva Fox Squirrel

Giant Kangaroo Rat

Blue Whale

Fin Whale

Humpback Whale

Bowhead Whale

Addax

Sonoran Pronghorn

Scimitar-Horned Oryx

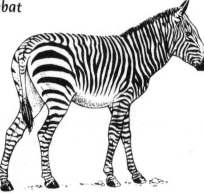

Birds

Jackass Penguin
Whooping Crane
White-Naped Crane
Eskimo Curlew
Hawaiian Goose (Nene)
California Condor
Harpy Eagle
Monkey-Eating Eagle
Florida Everglades Kite
Northern Aplomado Falcon
Attwater's Prairie Chicken
Kakapo
Bali Myna
Resplendent Quetzal
Golden-Cheeked Warbler
Black-Capped Vireo
Southwestern Willow Flycatcher
Red-Cockaded Woodpecker
Crested Honeycreeper
Micronesian Kingfisher
Great Indian Bustard

Reptiles

Galápagos Giant Tortoise
Hawksbill Turtle
Kemp's Ridley Sea Turtle
Leatherback Sea Turtle
Dumeril's Ground Boa
Tuatara
Galápagos Marine Iguana
St. Croix Ground Lizard
Gila Monster
Komodo Dragon
American Crocodile
Nile Crocodile

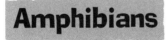

Fish

Paddlefish
Gila Trout
Bonytail Chub
Giant Catfish
Leon Springs Pupfish
Comanche Springs Pupfish
Big Bend Gambusia
Colorado River Squawfish
Cui-ui

Amphibians

Axolotl
San Marcos Salamander
Texas Blind Salamander
Houston Toad

Have you ever heard of mountain lions? What about cougars? Pumas? Catamounts? Panthers? All of these names describe the same species of large cat that has the scientific name *Felis concolor*. Why does this big cat have so many different common names? The reason is that these predators are found all over North and South America, and people in different places have called these cats different names.

Across their vast range, these cats live in a variety of habitats—from moist forests and cool mountains to steamy swamps. In some areas, small populations of these cats have become isolated from the general population. And over time, they have developed traits that are different enough from the general population to cause scientists to classify them as a subspecies.

Stalking the swamps and forests at the southern tip of Florida is one of these subspecies, the Florida panther, which has the scientific name of *Felis concolor coryi*. Before European settlers came to North America, Florida panthers were found throughout the southeastern United States. These sleek, brown cats were once the top predator in the Florida ecosystem, feeding on deer and other large prey. Today only 30 to 50 of them are left in the wild.

Florida panther

©Daniel J. Cox/Tony Stone Images

What happened?

During the day, panthers prefer to rest and sleep in areas that are higher and drier than the areas they use for hunting. People also prefer these drier areas for agriculture, pasture land, and housing developments.

A

Panthers were once considered a threat to people and farm animals. In 1887, the state of Florida started paying hunters $5 for every panther killed. Unlimited hunting of panthers wasn't stopped until 1958.

B

Panthers need a lot of space, especially male panthers. Individual males may have a territory of 200 square miles each, and they don't like to share it with other male panthers.

C

Between 1979 and 1991, collisions with cars caused 46 percent of panther deaths in Florida.

D

In the Everglades, panthers are being poisoned by mercury (a poisonous metal). At least one panther has died from mercury poisoning. Panthers probably become poisoned when they eat prey animals, such as raccoons, that have eaten poisoned food. Nobody is certain where the mercury is coming from, but scientists suspect it may be released into the air by burning trash or burning sugar cane fields, or it may be released when swamplands are drained for agriculture.

E

Melaleuca (pronounced mel-ah-LOO-kah) trees, Brazilian pepper shrubs, Australian pine, and other plant species that were introduced to Florida from other countries are growing like crazy and taking over much of the state's native habitat. Deer, an important part of the panther's diet, and the panthers themselves are having a hard time adapting to the new conditions.

F

Florida's human population has doubled every 20 years since 1830. The density of people has increased from less than 1 person per square mile in 1900 to 240 people per square mile in 1990.

G

Urban areas in Florida increased 650 percent between 1936 and 1990.

H

Windows on the Wild: Biodiversity Basics

World Wildlife Fund

As more people move into an area, they consume more resources, and this can affect panthers and other wildlife. When Florida residents water their lawns or fill their swimming pools, for instance, water gets channeled to urban areas instead of staying in natural areas.

I

Forty-three percent of Florida's native habitat has been changed to urban, suburban, and agricultural areas.

J

Water used to flow down to the Everglades from the middle of Florida, then spread through the Everglades like a 50-mile-wide river. Now much of that water flows through 1,000 miles of people-made canals and is used to irrigate crops and supply communities with drinking water.

K

Florida Panther Habitat

PANTHER SOLUTIONS

There are many ways people are trying to save the Florida panther. Here's a list of some of them.

- Federal and state laws, such as the federal Endangered Species Act and the Florida Panther Act of 1978, are designed to keep people from killing, disturbing, injuring, harming, or harassing panthers. People who violate these laws can wind up paying a large fine and may find themselves in jail.

- In 1989, the Florida Panther Wildlife Refuge was established in southern Florida. This special refuge protects 24,000 acres of panther habitat. The Big Cypress National Preserve and Fakahatchee Strand State Preserve, formed in 1974, also protect large areas of panther habitat.

- About 53 percent of the panther's range in Florida is on private land. Officials are working closely with private landowners to encourage them to preserve panther habitat on their land.

- Scientists captured wild panther kittens in 1991 to begin a captive breeding program. When these kittens grow up and have their own young, the young will be released into the wild. The ultimate goal is to have two new panther populations. The captive breeding program also helps scientists manage genetic diversity within the small Florida panther population. Another way scientists have tried to increase genetic diversity is by releasing cougars from Texas into southern Florida. Before the Florida panther became isolated in southern Florida, Texas cougars and Florida panthers interbred.

- Between 1981 and 1991, 43 panthers were fitted with special collars that helped scientists track the panthers' movements. The information gathered from this research has been used to protect the areas that panthers use and to begin the captive breeding program.

- People are working to restore the Everglades by undoing some of the changes people have brought to the area. For example, they are removing some of the canals so water can flow naturally through southern Florida just as it used to do. They are creating giant marshes near agricultural areas. These marshes will help to filter out pollutants from water that runs off agricultural fields before the water flows into the Everglades. There are also plans to buy back some agricultural lands and turn them into wildlife refuges.

- In 1982, the panther was adopted as the state animal of Florida. Florida residents can now buy a panther license plate for their car. The money raised from the license plate sales goes to education programs that help the public learn about the habitat needs of the Florida panther.

- Workers are digging up Melaleuca trees and other introduced plant species from areas throughout southern Florida. They are also setting controlled fires to kill introduced plants and to help the return of native plants that make up the panther's preferred habitat.

- Between 1979 and 1991 almost half the Florida panthers that died were killed by cars. However, when State Route 84 was converted to Interstate 75—one of the main roads running through panther country—36 special wildlife "tunnels" were built under it. No panthers have been killed by cars on Interstate 75 since the tunnels were built.

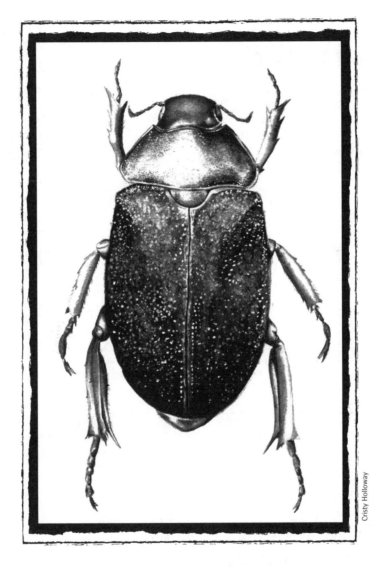

Cristy Holloway

*"Every time I see a beetle it blows
my mind. And working on tropical species, I get
personally excited when I see the tremendous variety
that's living with us on the planet."*

–Terry Erwin, entomologist

Name of Ecoregion:

Location:

Three or more species that live in this ecoregion:

1. _____

2. _____

3. _____

_____

Three interesting facts about this ecoregion or these species:

1. _____

2. _____

3. _____

This map highlights more than 200 of the richest, rarest, and most unique natural areas on the planet. From the windswept tundra of Alaska's North Slope to the warm tropical forests of the Congo Basin, these bio-rich areas are teeming with life.

Legend

Terrestrial Ecoregions

Marine Ecoregions

Freshwater Ecoregions

WWF

Okapi
[o-KA-pee] (Okapia johnstoni)

This hoofed mammal is related to giraffes. But it's so elusive that it wasn't named by scientists until the early 1900s. The okapi's front half is rich brown, and its rump and hind legs are striped like a zebra. This coloring helps it stay camouflaged in the Southern Congo Basin Forests of central Africa.

Bonobo
[bow-NO-bow] (Pan paniscus)

Bonobos, or "pygmy chimps," live deep in the forest, alongside such animals as okapis and forest elephants.

Salongo Monkey
[sa-LAWN-go] (Cercopithecus salongo)

This Old World monkey lives in the thicket of the moist broadleaf forest of central Africa. These monkeys spend a majority of their time on the ground foraging for fruits, leaves, and plant shoots.

Giant Baobab
[BAY-o-bab] (Adansonia grandidieri)

If you walked in the Madagascar Dry Forests, you'd be sure to see these tall, swollen trees. Like the rest of the plants in this region, baobabs have to survive a long and hot dry season. One way they do this is by having huge water-filled trunks, earning them the name of "bottle trees." Local people may tap a hole in the trunk of a baobab to obtain water for themselves or their livestock.

Sicklebilled Vanga
[VAN-ga] (Falculea palliata)

This beautiful black and white bird uses its long, curved bill to hook insects out of their holes. You might see one perched on the trunk of a giant baobab tree—searching for a juicy snack beneath a piece of bark.

Panther Chameleon
[ka-MEE-lee-on] (Chameleo pardalis)

Underneath giant baobabs grow scrubby bushes and trees where these chameleons hunt for insects. Panther chameleons are patient and observant hunters, with eyes that can swing fully in their sockets and look in two directions at once. As one of these chameleons slowly approaches its prey, it sways back and forth like a piece of vegetation moving in the wind. When the lizard is close enough, its sticky-tipped tongue flashes out and nabs the meal, bringing it back to the chameleon's mouth in a fraction of a second.

Wild Maize
[MAZE] (Zea diploperennis)

Monarch butterflies and imperial woodpeckers share their forests with a plant called wild maize. This wild grass is closely related to one of the world's most important food crops: corn. That means that genes from wild maize may be used in the future to protect domesticated corn from diseases and to help it adapt more easily to harsh environmental conditions.

Monarch Butterfly
[MAHN-arch] (Danaus plexippus)

Maybe you've heard of birds flying south for winter, but what about insects? In the fall, thousands of these brilliant black and orange insects fly from all parts of the eastern United States and Canada to the Mexican Pine-Oak Forests. Some monarchs travel as many as 3,000 miles! Then they spend the winter together in just a few isolated groves of trees.

Imperial Woodpecker
(Campephilus imperialis)

If you are very lucky, you may see this extremely threatened woodpecker. First you need to go to the pine trees where most monarch butterflies in North America spend their winters. Then look high in the branches where the imperial woodpecker often perches.

Cobra Lily
(Darlingtonia californica)

In open bogs surrounded by bushes of western azaleas, you may find these plants. They form large tubes that look like a giant hooded cobra with a forked tongue. Insects that crawl inside the hood to explore are trapped by bristles and fall into a pool of liquid where they are digested by the plant.

Western Azalea
[a-ZAY-lee-uh] (Rhododendron occidentale)

Delicate pink flowers decorate the western azalea. This plant grows in the shade of the Klamath-Siskiyou Coniferous Forests, which have the highest diversity of needle-leafed trees in the world. While walking in these forests, you might see the low-hanging branches of the Brewer's spruce tree or notice the sweet smell of the ponderosa pine.

Siskiyou Mountains Salamander
[SIS-ka-you] (Plethodon stormi)

If you were looking for these amphibians, you might find them crawling in a bog where cobra lilies grow. Or you might find them hiding in the cool pine needles and other tree debris underneath a western azalea.

Fountain Bamboo
[bam-BOO] (Fargesia nitida)

The fountain bamboo is a giant grass that grows in mountain forests. Some groves of fountain bamboo bloom only once every 100 years. And after they flower, they quickly die. If this happens over a large area, giant pandas and other animals that eat bamboo may have to travel great distances to find enough food to survive.

Windows on the Wild
BIODIVERSITY BASICS

Giant Panda
(Ailuropoda melanoleuca)

Wild giant pandas live only in the Central/Southwest China Temperate Forests. Among their favorite foods are the stems and leaves of fountain bamboo.

Windows on the Wild
BIODIVERSITY BASICS

Golden Pheasant
[FEZ-ant] (Chrysolophus pictus)

The fruits and seeds of various plants are the food of the golden pheasant. This rare bird lives in the same mixed forests where you might find a giant panda.

Windows on the Wild
BIODIVERSITY BASICS

Wolverine
[WOOL-ver-een] (Gulo gulo)

These creatures are small but fierce! They hunt small mammals in the cold forests of Siberian spruce that cover northern Asia.

Windows on the Wild
BIODIVERSITY BASICS

Long-Eared Owl
(Asio otus)

During the day, long-eared owls roost in the high branches of trees such as Siberian spruce. Their patchy brown color helps them blend in with their surroundings. At night, they hunt for rats, mice, rabbits, and other small mammals that share their cold forest home.

Windows on the Wild
BIODIVERSITY BASICS

Siberian Spruce
(Picea obovata)

The Siberian spruce is one of the common trees found in the largest unbroken wilderness in the world: the Central and Eastern Siberian Boreal Forests and Taiga.

Windows on the Wild
BIODIVERSITY BASICS

Norway Lemming
(Lemmus lemmus)

During the long, dark winter of the Arctic region, lemmings make tunnels under the snow as they search for grasses, shrubs, mosses, and other food. But every few years, these rodents experience a population explosion, and large groups of them move above ground into open fields and valleys, which they share with large herds of reindeer. Contrary to popular belief, the lemmings do not migrate to the sea and commit suicide, although most of those that wander into the open are eaten by predators.

Windows on the Wild
BIODIVERSITY BASICS

Reindeer
(Rangifer tarandus)

The reindeer (or caribou) is the only deer in which both sexes have antlers. During the long days of summer, they feed on grass and other plants in the tundra. In the winter, they must scrape away snow to expose dry lichens, which can endure in the intense cold. In the Scandinavian Alpine Tundra and Taiga, local people such as the Sami depend on reindeer for meat, milk, and clothing.

Windows on the Wild
BIODIVERSITY BASICS

Arctic Fox
(Alopex lagopus)

The arctic fox has fur-covered feet and small, rounded ears to reduce heat loss in the cold tundra. Arctic foxes don't hibernate in winter. To track down a meal when times are lean, they may listen for the sounds of lemmings and other small mammals scrambling under the snow.

Prairie Dog-Tooth Violet
(Erythronium mesochoreum)

The small white flowers of the prairie dog-tooth violet are among the first signs of spring in the vast Tallgrass Prairies of the central United States.

Greater Prairie Chicken
(Tympanuchus cupido)

As you wade through a sea of tall grass, you may hear the booming call of the male greater prairie chicken. The male birds perform spectacular courtship displays—inflating large pouches on their necks, stamping their feet, and fanning their tail feathers to attract females. They share their home with a variety of plants and animals, from the tiny prairie dog-tooth violet to the enormous bison.

American Bison
(Bison bison)

Just 200 years ago there were millions of bison roaming the grasslands of North America. By the beginning of the 20th century, they were almost extinct. Now bison are increasing, thanks to protection efforts. But most of their vast grassland home has become agri-cultural land. Greater prairie chickens, prairie dogs, wolves, and rattlesnakes are a few of the animals that live side by side with the bison in the wild.

One-Horned Rhinoceros
(Rhinoceros unicornis)

The one-horned rhinoceros grazes in fields of grass that grows over 20 feet tall! The rhinos' hides are made of thick plates of skin with deep folds that resemble a suit of armor. The thick hides protect the rhinos from the largest predators of their habitat—tigers. Other animals that share these grasslands include Asian elephants, chital, and Indian bison.

Windows on the Wild
BIODIVERSITY BASICS

Chital
[CHEED-al] (Axis axis)

This deer has a rich brown coat speckled with white spots, which help to camouflage it in the high grass of the Terai-Duar Savannas and Grasslands of central Asia. It has an acute sense of hearing and will dash away at the slightest hint that its main predator, the tiger, is nearby.

Windows on the Wild
BIODIVERSITY BASICS

Tiger
(Panthera tigris)

Tiger stripes are more than pretty decoration. The stripes help tigers blend in with light and shadows as they stalk wary prey, such as chital, in the high grass. Unfortunately, people have hunted these rare cats to near extinction because of their handsome coats and because many other parts of their bodies are believed to have medicinal value.

Windows on the Wild
BIODIVERSITY BASICS

Snail Kite
(Rhostrhamus sociabilis)

In order to make room for houses, shopping centers, and farms, people have drained the water from the snail kite's habitat. Unfortunately, this change has drastically reduced populations of the bird's main food source: apple snails.

Windows on the Wild
BIODIVERSITY BASICS

Florida Tree Snail
(Liguus fasciatus)

Years ago, a visitor to the Everglades Flooded Grasslands might have found hundreds, or even thousands, of shells belonging to the Florida tree snail. But as the Everglades have been drained for housing developments and agriculture, Florida tree snails and many other kinds of snails (such as the apple snail) are becoming rare. And now the animals that depend on the snails for food are in danger of becoming extinct.

Windows on the Wild
BIODIVERSITY BASICS

Red Bay
(Persea borbonia)

Clumps of red bay trees rise like islands in the sea of sawgrass that covers their flooded homes. Almost two dozen different species of tree snails, including the Florida tree snail, live and feed on red bays. But the snails don't eat the trees' leaves. Instead, they <u>n</u>ibble on algae that grow on the red bays' bark.

Windows on the Wild
BIODIVERSITY BASICS

Andean Condor
[AN-dee-an KON-door] (Vultur gryphus)

These condors are the largest flying birds on <u>E</u>arth—soaring at altitudes of up to 20,000 feet. They feast on the dead carcasses of vicuña, llama, alpaca, and other large grazing animals that live in the cold mountain heights.

Windows on the Wild
BIODIVERSITY BASICS

Frailejón
[Fray-eelay-HONE] (Espeletia spp.)

The fuzzy gray leaves of the frailejón are perfectly adapted to the high altitudes of the North Andean Paramo. The leaves resist the harsh ultraviolet rays of the sun and can tolerate the region's freezing nights. Look up and you may see an Andean condor soaring near the tops of the glacier-covered peaks that loom in the distance.

Windows on the Wild
BIODIVERSITY BASICS

Vicuña
[vi-KOON-ya] (Vicugna vicugna)

The silky cinnamon-colored fur of the vicuña was so highly valued in ancient times that only the royal people of the Inca (an empire that flourished in the 1500s) were allowed to use it. These animals roam the high mountains, wandering among the fuzzy frailejón and other plants. The small vicuña is related to domesticated llamas and alpacas, animals important to local people for their meat and wool.

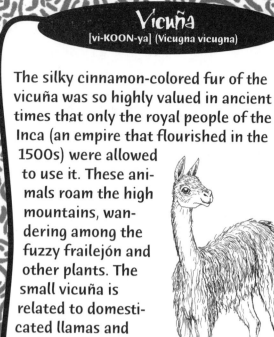

Welwitschia Plant
[wel-WIT-shee-a] (Welwitschia mirabilis)

This unusual plant produces only two leaves, which can grow for the entire life of the plant. And that can be more than 1,500 years! The welwitschia plant grows in the extreme conditions of the Namib and Karoo Deserts and Shrublands of southern Africa, where it may not rain for more than five years in a row.

Darkling Beetle
(Onymacris unguicularis)

These beetles feed on dead animal and plant material that collects in their rugged desert home. Though there is almost no rain, these beetles have developed ways to obtain water from the fog that commonly settles around the dunes. Some tip their abdomens upright so that the condensation on their backs runs into their mouths. Others dig trenches to collect fog water on the ground. During the heat of the day, these beetles may seek shelter underneath the sand or in the shade of a welwitschia plant.

Sand-Diving Lizard
(Aporosaura anchietae)

Sand-diving lizards hunt for darkling beetles, spiders, scorpions, and other small animals that scuttle across the desert. When things get hot, the lizards dive under the sand or seek out the shade of plants, such as the welwitschia. In doing so, they help reduce water loss from their bodies.

Numbat
[NUM-bat] (Myrmecobius fasciatus)

With its elegant striped coat, the numbat is one of the most beautiful marsupials in the world. Unlike most marsupials, numbats are active during the day. They use their thick claws to open termite nests, lapping up the insects with their long tongues. Some of their main predators in the wild are hunting birds such as little eagles and brown goshawks.

Windows on the Wild
BIODIVERSITY BASICS

Brown Goshawk
[GOSS-hawk] (Accipiter fasciatus)

The brown goshawk makes its home in the Southwest Australian Shrublands and Woodlands. There the bird uses its sharp eyes to hunt down honey possums, numbats, and other small mammals.

Windows on the Wild
BIODIVERSITY BASICS

Drosera Sundew
[DROSS-er-a] (Drosera menziesii)

As numbats nab termites with their long tongues, meat-eating sundew plants are using their own sticky trick to catch a meal. Passing insects are attracted to the pink flowers on these lovely plants. But when they land on the leaves, they get stuck! The plant's dew is a super sticky substance. The globs of "glue" are on the ends of hairs that slowly curl around the insect and digest it. In the process, the plant absorbs valuable <u>nutrients</u>.

Windows on the Wild
BIODIVERSITY BASICS

Platypus
[PLAT-i-puss] (Ornithorhynchus anatinus)

It has a beak and webbed feet, but it isn't a bird! The platypus is actually a "duck-billed," egg-laying mammal that lives in Eastern Australian Rivers and Streams. The platypus pokes around the bottom of the river with its flexible bill to find the small animals it eats. But it keeps its eyes, ears, and nostrils closed—using only touch receptors to find its way under water.

Windows on the Wild
BIODIVERSITY BASICS

Freckled Duck
(Stictonetta naevosa)

During the day, the rare freckled duck hides out in dense vegetation to avoid the Murray cod and other predators. Only at night does it venture out into shallow waters to feed on plankton and algae. With its head below the surface, it may encounter other water creatures, such as the platypus, that share the rivers it calls home.

Windows on the Wild
BIODIVERSITY BASICS

Murray Cod
(Maccullochella peeli)

Swimming in the same waters as the platypus is this enormous and incredibly aggressive fish. Weighing about 255 pounds, the Murray cod preys on almost anything in its path, including the rare freckled duck, other waterfowl, and rats.

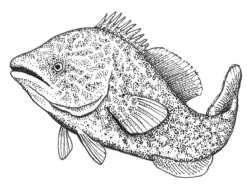

Windows on the Wild
BIODIVERSITY BASICS

Amazon Dolphin
(Inia geoffrensis)

This freshwater dolphin spends its days avoiding fishing nets and looking for food. It feeds on fish that eat fruit that drops into the water from surrounding trees. Because of its pink color, this dolphin may be the source of many legends in the Amazon that tell of an animal that can change from human to dolphin.

Windows on the Wild
BIODIVERSITY BASICS

Arawana Fish
[arr-a-WA-na] (Osteoglossum bicirrhosum)

Arawanas live in the Varzea and Igapó Freshwater Ecosystem and can grow to be about three feet long. They have been known to jump several feet out of the water to catch small birds for food! They also feed on insects that land on the surface of the water. Arawanas have a rough tongue covered with sharp, tooth-like projections that trap food up against the roof of their mouths.

Windows on the Wild
BIODIVERSITY BASICS

Black Caiman
[KAY-man] (Melanosuchus niger)

If you visit the Amazon, watch out for a head sticking out of the water with yellow or white bands on the lower jaw. Black caimans are crocodilians that feed on fish that swim in the freshwater rivers of the Amazon Basin. But as they get older, they may come ashore to feed on rodents, domestic animals, and sometimes even humans.

Windows on the Wild
BIODIVERSITY BASICS

Nerpa
[NER-pa] (Phoca sibirica)

It is a mystery how this seal, known as the nerpa, made its way to Lake Baikal—which is hundreds of miles from any sea or ocean! The only freshwater seal in the world, the nerpa's warm pelts and fat have made it a favorite of hunters for thousands of years.

Windows on the Wild
BIODIVERSITY BASICS

Golomyanka
[go-low-mee-ON-ka] (Comephorus baicalensis)

This fish, known as the golomyanka, can be found swimming in the icy waters of Lake Baikal. This transparent fish makes a great meal for hungry nerpa seals. The golomyanka also shares its harsh lake home with tiny filter-feeding crayfish.

Windows on the Wild
BIODIVERSITY BASICS

Gammarid Shrimp
[ga-MARR-id] (Eulimnogammarus spp.)

Known as caretakers of their icy lake home, these tiny crayfish devour everything that threatens to pollute their cold, watery habitat, including dead fish and insects. They ensure that their golomyanka and nerpa neighbors have a clean home.

Windows on the Wild
BIODIVERSITY BASICS

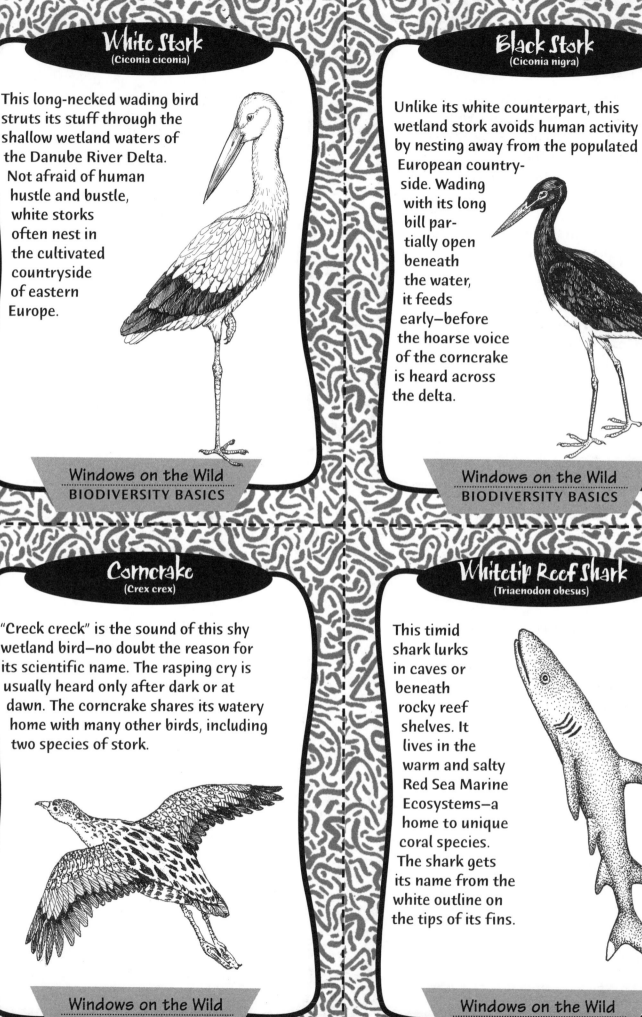

White Stork
(Ciconia ciconia)

This long-necked wading bird struts its stuff through the shallow wetland waters of the Danube River Delta. Not afraid of human hustle and bustle, white storks often nest in the cultivated countryside of eastern Europe.

Windows on the Wild
BIODIVERSITY BASICS

Black Stork
(Ciconia nigra)

Unlike its white counterpart, this wetland stork avoids human activity by nesting away from the populated European country-side. Wading with its long bill par-tially open beneath the water, it feeds early—before the hoarse voice of the corncrake is heard across the delta.

Windows on the Wild
BIODIVERSITY BASICS

Corncrake
(Crex crex)

"Creck creck" is the sound of this shy wetland bird—no doubt the reason for its scientific name. The rasping cry is usually heard only after dark or at dawn. The corncrake shares its watery home with many other birds, including two species of stork.

Windows on the Wild
BIODIVERSITY BASICS

Whitetip Reef Shark
(Triaenodon obesus)

This timid shark lurks in caves or beneath rocky reef shelves. It lives in the warm and salty Red Sea Marine Ecosystems—a home to unique coral species. The shark gets its name from the white outline on the tips of its fins.

Windows on the Wild
BIODIVERSITY BASICS

Giant Clam
(Tridacna gigas)

Like its white-tipped neighbor and the sea in which it lives, the giant clam is known for its coloring. Sparkling blues and greens on the clam's mantle (the membrane between the body and the shell) actually result from colorful algae that live in the exposed tissue. These huge mollusks can sometimes live for 50 years and grow to be several hundred pounds. They share the reef waters with green turtles.

Windows on the Wild
BIODIVERSITY BASICS

Green Turtle
(Chelonia mydas)

Grazing on sea grasses and algae of the reef, green turtles are known as the cattle of the warm seas. They normally live in reef waters alongside whitetip reef sharks and giant clams. But they may also migrate through hundreds of miles of open ocean to lay their eggs.

Windows on the Wild
BIODIVERSITY BASICS

Puffin
(Fratercula arctica)

Known as the "sea parrot," this unusual bird uses its wings to power its penguin-like body through the waters of Icelandic and Celtic Marine Ecosystems. As they dive down to 150 feet, these puffins use their feet to steer. They can carry as many as 28 small fish in their beaks at once!

Windows on the Wild
BIODIVERSITY BASICS

Narwhal
[NAR-wall] (Monodon monoceros)

When sailors in the 1700s came home carrying samples of the narwhal's spiraling tusk, people thought the tusks were proof that unicorns existed. A narwhal's tusk is strange, but it's really nothing more than a modified tooth. Male narwhals may use their tusks as a weapon in their fights for female attention. The deep-diving puffin and gray seal may be nearby, but they'll be minding their own business.

Windows on the Wild
BIODIVERSITY BASICS

Gray Seal
(Halichoerus grypus)

Gray and brown fur helps keep these seals warm in their icy home waters. They may snack on some of the same fish that puffins in the area eat. Large groups of seals may gather on secluded beaches, but individuals keep their distance. They don't have a weapon anything like their narwhal neighbors do, but they still have a tendency to get into fights with other seals!

Windows on the Wild
BIODIVERSITY BASICS

Walrus
(Odobenus rosmarus)

The polar region around the Bering and Beaufort Seas is home to herds of these highly social creatures. Walruses have long tusks that are used in dominance displays and sometimes as weapons. Between the tusks is a snout full of sensitive whiskers that help the walrus locate a meal in murky waters.

Windows on the Wild
BIODIVERSITY BASICS

Polar Bear
(Ursus maritimus)

Lying on their backs with their feet in the air, these powerful carnivores often nap on polar ice after a hunt. Polar bears aren't picky eaters—if there are no live seals in the area, the carcass of a bowhead whale or a walrus will do for dinner!

Windows on the Wild
BIODIVERSITY BASICS

Bowhead Whale
[BOE-head] (Balaena mysticetus)

The heads of these whales make up one-third of their total length. And their arching jaws give them their "bowed" shape. Bowhead whales were at one time endangered because of commercial whaling, and their numbers remain low. Still, polar bears and walruses may see a female bowhead and her calf swimming offshore, or they may encounter a three-ton carcass washed up on the beach.

Windows on the Wild
BIODIVERSITY BASICS

SECRET MESSAGE CARDS

The island of Kauai (part of the Pacific Ocean's Hawaiian Island chain) has more of this than any other place on Earth.

___ ___ ___ ___

To find the answer, look in the Namib Desert, the Scandinavian Alpine Tundra, the Klamath-Siskiyou Coniferous Forests, and the Southwest Australian Shrublands and Woodlands.

This phenomenon is created by the gravitational pull of the sun and moon on planet Earth and is an important force that constantly changes the shape of the land's surface.

___ ___ ___ ___

To find the answer, go to the Mexican Pine-Oak Forests, the Central/Southwest China Temperate Forests, the Klamath-Siskiyou Coniferous Forests, and the Central and Eastern Siberian Boreal Forests.

In the deciduous forests of the eastern United States, the elimination of large predators and the increase of edge habitat has led to a dramatic increase in the population of these animals.

___ ___ ___ ___

To find the answer, look in the Madagascar Dry Forests, the Tallgrass Prairies of the United States, the Mexican Pine-Oak Forests, and the Central and Eastern Siberian Boreal Forests.

In the summer, you may find one of these normally seafaring birds hundreds of miles from the ocean, breeding in vast colonies in places such as the high plateaus of Tibet in central Asia.

___ ___ ___ ___

To find the answer, head to the Scandinavian Alpine Tundra, the North Andean Paramo, the Central/Southwest China Temperate Forests, and the Southwest Australian Shrublands and Woodlands.

In many areas of the world, the establishment of one of these can lead to vast amounts of pollution as chemicals and sediments wash into nearby watersheds.

___ ___ ___ ___

To find the answer, look in the North Andean Paramo, the Southern Congo Basin Forests, the Klamath-Siskiyou Coniferous Forests, and the Everglades Flooded Grasslands.

When people do this to wetlands to build houses or create agricultural land, many valuable services the wetlands provide (such as controlling floods and serving as a nursery for sea life) are lost.

___ ___ ___ ___ ___

To find the answer, head to the Klamath-Siskiyou Forests, the Tallgrass Prairies of the United States, the Terai-Duar Savannas and Grasslands, the Mexican Pine-Oak Forests, and the Everglades Flooded Grasslands.

A portion of soil the same size as this object might contain millions of microorganisms, including fungi, bacteria, and tiny animals.

___ ___ ___ ___

To find the answer, look in the Klamath-Siskiyou Coniferous Forests, the Madagascar Dry Forests, the Southern Congo Basin Forests, and the Mexican Pine-Oak Forests.

Bonus Question

This habitat type holds 20 percent of the Earth's plant species, and every ecoregion within it is in critical danger of becoming extinct. What is it? (The answer uses all the letters highlighted in the Ecoregion Species Cards.)

___ ___ ___ ___ ___ ___ ___ ___ ___ ___ ___ ___

How many ecoregions are there within this habitat type?
Does this habitat type exist in the United States? If so, where?

138

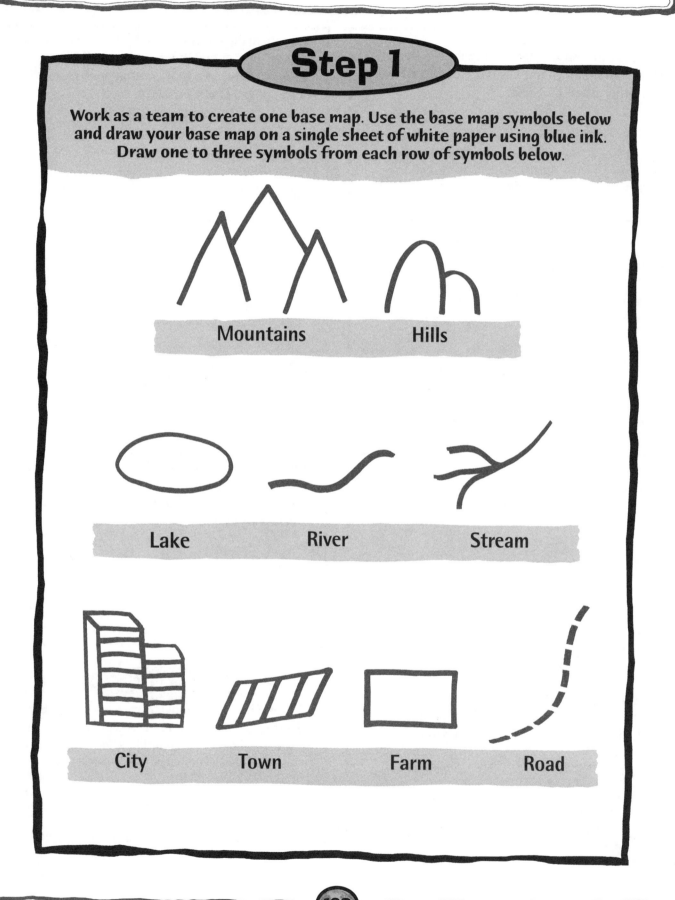

Step 1

Work as a team to create one base map. Use the base map symbols below and draw your base map on a single sheet of white paper using blue ink. Draw one to three symbols from each row of symbols below.

Mountains   Hills

Lake   River   Stream

City   Town   Farm   Road

Step 2

Have each person on your team choose one of the categories below (ecosystems of special concern, rare species, or protected areas). Then, one at a time, have each person place a blank transparency over the base map and draw in the symbols for his or her category. Remember to follow the directions next to each category and not to let others in the group see where you've drawn your symbols.

ECOSYSTEMS OF SPECIAL CONCERN

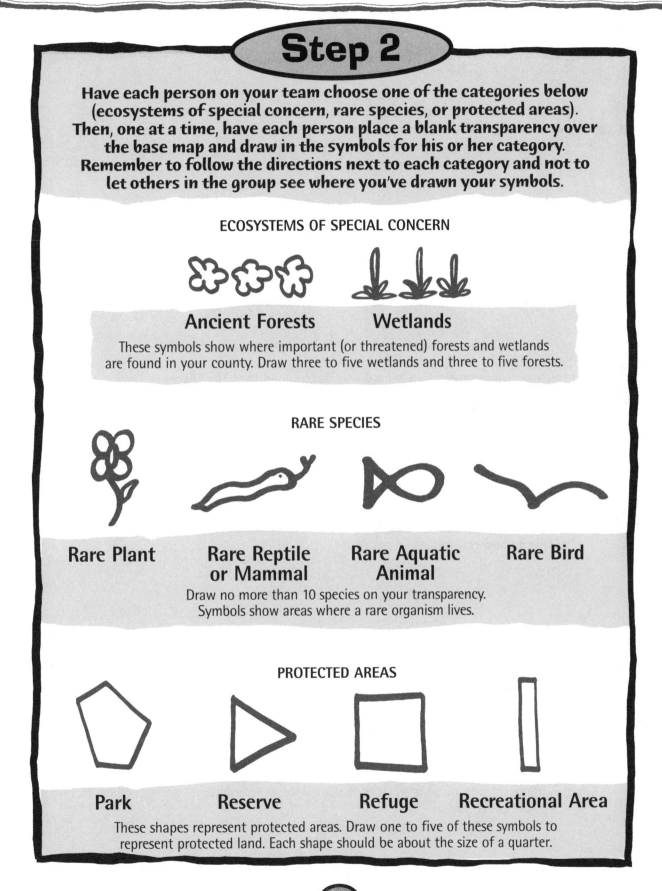

Ancient Forests   **Wetlands**

These symbols show where important (or threatened) forests and wetlands are found in your county. Draw three to five wetlands and three to five forests.

RARE SPECIES

Rare Plant   **Rare Reptile or Mammal**   **Rare Aquatic Animal**   **Rare Bird**

Draw no more than 10 species on your transparency. Symbols show areas where a rare organism lives.

PROTECTED AREAS

Park   **Reserve**   **Refuge**   **Recreational Area**

These shapes represent protected areas. Draw one to five of these symbols to represent protected land. Each shape should be about the size of a quarter.

World Wildlife Fund   *What's the Status of Biodiversity?*

1. Which rare species or ecosystems of special concern are already protected in your county? Where are the gaps in protection?

2. Make a list of some of the threats that could endanger the rare species and ecosystems of concern that lie outside the protected areas.

3. Using your gap analysis, how would you recommend that The Diversity Trust, other conservation groups, or the county planning department help protect the rare species or ecosystems in Wild Hills County?

4. Besides buying or protecting land, what are some other ways to protect rare species and ecosystems of concern?

5. What would you do if you discovered the following information?

- One of your "rare" species has just been found in abundant numbers in an adjacent county.

- A new factory has just been built in one of the towns.

- A national report has rated your county as the top place in the country to live. Developers are expanding suburban areas to accommodate new families and businesses.

- A major portion of your wetland is owned by an elderly person very dedicated to wetland preservation.

HOW MUCH IS A GRAY WOLF WORTH?

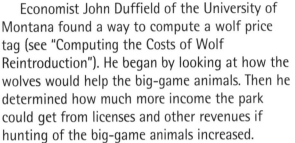

Suppose you were trying to convince officials to reintroduce wolves into Yellowstone National Park. You might try appealing to their sense of history—reminding them that lots of gray wolves roamed the region until the federal government paid people to kill off the wolves during the 1930s.

Or perhaps you'd try an ecological argument—pointing out how gray wolves help limit the large populations of bison and elk that have overgrazed the park's vegetation.

You might even try an emotional plea—saying that people will have a more meaningful visit to Yellowstone if they hear wolves howl. But can you imagine arguing that it's more *profitable* to have wolves in Yellowstone National Park?

That's just what wolf-lovers tried several years ago. Park officials weren't showing any signs of wanting to bring back the wolves. So the wolf-supporters started to argue that the financial benefits of having wolves in Yellowstone far outweighed the costs.

But how do you put a price tag on a wolf? It wouldn't be hard to fix a price on a product such as a wolf photograph or T-shirt. It's not even hard to imagine coming up with a price for a pet or for a wolf that's being sold to a zoo. But what's the price of a wolf roaming free in a national park? Surely it's worth something, but if nobody buys it, how do we know how much it's worth?

Economist John Duffield of the University of Montana found a way to compute a wolf price tag (see "Computing the Costs of Wolf Reintroduction"). He began by looking at how the wolves would help the big-game animals. Then he determined how much more income the park could get from licenses and other revenues if hunting of the big-game animals increased.

Next, Duffield found out how much longer visitors said they would stay in Yellowstone, and how much more often they'd go, if they knew they could hear wolves howl. He added the total of all those extra tourist dollars to the hunting revenues.

Finally, he subtracted the cost of livestock that the wolves might kill in areas around the park.

The bottom line? Bringing wolves back to Yellowstone would generate $18 million in extra income for the local economy in the first year, and about $110 million over 20 years.

It's important to realize why this wolf price tag was so important. In many environmental controversies, critics say that conservation doesn't

©Erwin and Peggy Bauer/Bruce Coleman Inc.

make economic sense. In large part, that's because wildlife and other natural resources are considered to be free. So if you have a choice between saving trees ($0) and cutting them down for sale (lots of dollars), cutting them down will always win the economic side of the argument.

But are those living trees really worth nothing? Of course not, say natural resource economists. They provide wildlife habitat, cleaner air, more stable soils, and scenic beauty, to name just a few benefits. So if you can find a way to put a price on those services, you can more accurately represent the benefits of "free" resources.

But there is still a lot of controversy surrounding natural resource economics. Critics are especially concerned about how economists come up with the price tags for these goods.

Some methods are fairly straightforward. To figure out how much a park is valued by its visitors, for example, economists simply add up all the entrance fees of visitors and find out what they are willing to pay to travel there.

But things get fuzzier when economists try to fix a price on something more abstract—especially resources we *don't* want to use. For example, how much is it worth to you to know that elephants exist? Or that the Grand Canyon isn't spoiled by smog? Or that your children will be able to play in woods near your home?

In these cases, economists try to get their numbers through surveys of the public. In Colorado, economists asked residents how much they'd be willing to pay to have a polluted local river cleaned up. Those near the site said they'd pay an average of $73 each; households in other parts of the state said they'd pay an average of $3.90, bringing the statewide total to $15.7 million.

As you might guess, these kinds of surveys come under attack because it's not clear that people *would* pay what they say they'd pay. So to further gauge the value of cleaning up that Colorado river, the economists tried another approach. They added up the estimated loss in property value that the polluted river had caused for the 500 houses closest to it. The results were an average of $24,400 each, or $12.2 million total.

In the Colorado river example, as in the case of the wolf, the numbers helped officials weigh the costs and benefits of restoring a natural resource. The result? In 1983 Colorado won a lawsuit from the polluters of the river and is using the money to clean it up. And in 1995 wolves were successfully reintroduced to Yellowstone National Park.

Of course, using natural resource economics doesn't guarantee that every conservation project should or will be carried out. But perhaps it will provide a fairer way of assessing the value of wildlife and wild places at a time when everyone is concerned about the bottom line.

Adapted from "How Much Is a Gray Wolf Worth?" by Ellen Brandt. *National Wildlife* (June/July 1993): 4–13.

COMPUTING THE COSTS OF WOLF REINTRODUCTION

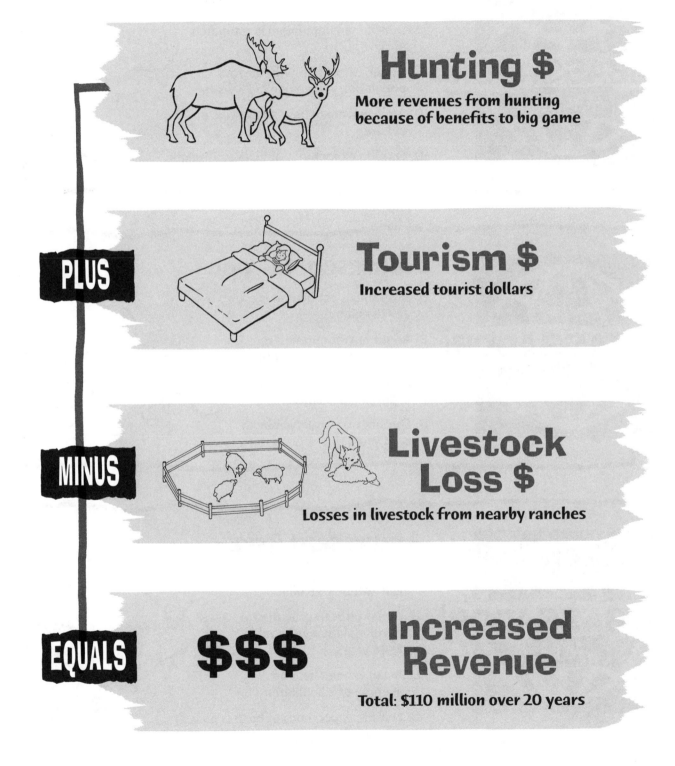

Hunting $
More revenues from hunting because of benefits to big game

PLUS

Tourism $
Increased tourist dollars

MINUS

Livestock Loss $
Losses in livestock from nearby ranches

EQUALS **$$$** **Increased Revenue**
Total: $110 million over 20 years

I represent Africa.

A. Our human population is estimated at 763 million.

B. At our current growth rate, our population will double in 27 years.

C. African women bear an average of 5.6 children.

D. Our life expectancy at birth is 52 years.

I represent Europe.

A. Our human population is estimated at 728 million.

B. At our current growth rate, our population will not double.

C. European women bear an average of 1.4 children.

D. Our life expectancy at birth is 73 years.

I represent Asia.

A. Our human population is estimated at 3.6 billion.

B. At our current growth rate, our population will double in 46 years.

C. Asian women bear an average of 2.8 children.

D. Our life expectancy at birth is 65 years.

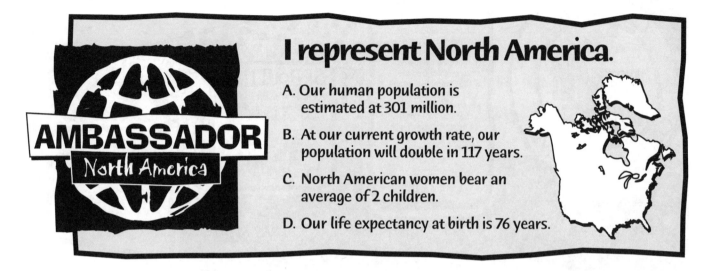

I represent North America.

A. Our human population is estimated at 301 million.

B. At our current growth rate, our population will double in 117 years.

C. North American women bear an average of 2 children.

D. Our life expectancy at birth is 76 years.

AMBASSADOR
North America

I represent Latin America.

A. Our human population is estimated at 500 million.

B. At our current growth rate, our population will double in 38 years.

C. Latin American women bear an average of 3 children.

D. Our life expectancy at birth is 69 years.

AMBASSADOR
Latin America

ORGANIC COTTON FARM

PLANTING

Fertilize with organic compost and manure, namely cow and chicken poop. Crop rotation and erosion control conserve topsoil; a nearby cover crop like alfalfa controls weeds and pests. Since healthy land will produce hardy, resistant plants, the seeds have not been pre-treated with chemicals that prevent plant diseases (fungicides).

SPROUTING

Not much to do but wait until the seedlings pop up. Non-treated seed sprouts faster.

ONCE THE PLANTS ARE UP

Spend lots of time with the young plants. Cultivate, cut the weeds with a tractor, bring in a crew to chop weeds by hand.

A TALE OF TWO T-SHIRTS

Two T-shirts side by side.... one made from organic cotton; the other made from cotton grown by conventional methods. They look so much alike, you wouldn't guess that their shirt tales are not the same. Here are the life stories of two tees, from field to fiber, on farms that grow cotton in very different ways.

SPROUTING

Much the same except that treated seed will take a bit more time to sprout.

ONCE THE PLANTS ARE UP

Bring in more herbicides and a crew to hand-chop the weeds.

CONVENTIONAL COTTON FARM

PLANTING

Most of the seeds have been pre-treated with fungicides. For bigger yields, farmers often sterilize the soil, replacing natural nutrients with synthetic fertilizers. Before the seed emerges, they also add chemicals that kill or control weeds (herbicides).*

HAZARDOUS

AS THE PLANT MATURES

Cultivate 2-3 more times. Irrigate. Organic cotton requires less irrigation (1½ to 3 acre-feet) because the soil holds more water. The beginning dressing of compost fertilizer is enough to take care of the whole season.

INSECT CONTROL
(Lygus, Spider Mites, Aphids & Worms)

Use beneficial bugs, the natural enemies of cotton pests. (They may already be present because insecticides haven't killed the good bugs along with the bad bugs.) Walk the field often and apply predators like praying mantids, lady beetles and lacewings to "hot spots" where pests might lurk. This takes patience.

LETTING THE PLANT MATURE

Irrigation stops and the plants mature. The cotton opens up in 45-60 days. Instead of removing the leaves (defoliating), plants die a natural death.

GINNING

Once cotton is harvested, the gin separates seed, leaf and trash from the cotton fiber. Before organic cotton is processed, the gin is cleared of residue from conventional cotton. The organic cotton is baled, tagged and kept separate from any conventional cotton. Then it's off to its future life as a T-shirt.

COTTON

AS THE PLANT MATURES

The plants need a nutrient boost. Side-dress with synthetic fertilizer. Water use will average 2-4 acre-feet.

INSECT CONTROL
(Lygus, Spider Mites, Aphids & Worms)

Some of the most serious chemicals are now applied by air or ground rigs. California alone uses 6,000 tons of pesticides on cotton in a single year.

LETTING THE PLANT MATURE

Farmers may use chemical regulators to slow plant growth and chemical defoliants to remove the leaves.

⚠ WARNING

GINNING

The gin is not cleaned and the cotton is not separated out.

*Many of the chemicals used in this and other phases of conventional cotton farming adversely affect human health and are toxic to birds, fish, amphibians and aquatic insects. For more information about the chemicals used to grow conventional cotton, see:
Basic Guide to Pesticides: Their Characteristics and Hazards
Shirley Briggs/Rachel Carson Council
©1992 Taylor & Francis Publishers, Washington, DC and London, UK

Reprinted with permission of Patagonia, Inc.

by Daniel Imhoff

1. The Cotton Crisis

Daniel Imhoff is a journalist based in San Francisco, California, who writes about issues of design, the environment, and agriculture. Mr. Imhoff helped develop one of the first organic cotton clothing lines at Esprit. He maintains a Web site on organic cotton.

Most people probably don't associate my home state of California with cotton farming. Sunny beaches, Hollywood, the Golden Gate Bridge, giant redwoods—maybe. But not one million acres of cotton, an area one and a half times the size of Rhode Island. Most people probably don't associate cotton farming with an excess of pesticides and other farm chemicals either. Yet cotton farming is responsible for a huge percentage of the world's use of agricultural chemicals, and environmentalists and others around the world are becoming increasingly alarmed about it. In California alone, over 17.5 million pounds of chemicals were applied to 1.1 million acres of cotton in 1995. That's nearly 16 pounds per acre (not including synthetic fertilizers). Each autumn, residents in California's cotton belt crowd into doctors' offices complaining of respiratory illnesses and flu-like symptoms. Migratory birds that succeed in reproducing are experiencing birth defects in record numbers.

Most farmers want to grow the finest crops possible, with the highest yields per acre, while minimizing the costs of growing. And they want to sell their harvest at a maximum profit. Yet how farmers achieve those goals depends on their own personal philosophy and methods. Two methods used for growing cotton are *conventional* farming and *organic* farming.

A conventional farmer relies upon a continual succession of inputs (chemicals, fuel, and machinery) to create ideal conditions for plants to thrive. For problems at every stage of the growing process there is a chemical solution. Fungicides protect the seedlings from fungus in the soil. Herbicides kill rival weeds. Pesticides kill insects. Defoliants strip the leaves off the plant and expose the cotton bolls (the white fuzzy part) to the whirling spindles of the picking machines. Most growers apply these chemicals with caution. That's because the chemicals are expensive and often very toxic. But by the end of the season, the quantities the farmers spread add up. Some of these chemicals persist long after they've been sprayed. They can adversely affect farm families, the surrounding communities, and wildlife living nearby. Still, the conventional farm industry—supported by chemical companies, government agencies, pesticide control advisors, money lenders, and university extension departments—insists that cotton cannot be grown economically without these chemicals.

1. The Cotton Crisis (Cont'd.)

In the early 1990s, however, a small number of farmers began applying the principles of organic farming to large-scale cotton cultivation. The results have been encouraging. These growers differ from their conventional counterparts in a few fundamental ways, starting with how they view the soil. Organic farmers believe that healthy, vigorous plants depend on living soil. They add natural fertilizers, such as manure, compost, and nitrogen-rich cover crops, to create nutritious soil. Under the best conditions, their organic soil becomes home to earthworms and diverse populations of microorganisms, and provides a steady stream of nutrients to the cotton plants. It takes three years of non-chemical farming for the land to be certified as organic. During those three years the farmer's crops are labeled "transitional organic."

An organic farmer considers the field, both above and below the soil, to be a living ecosystem that reaches a balance over time. The farm is not a sterile environment that is controlled by chemicals, but a living system that is maintained by a creative interaction with nature. For example, organic growers encourage diverse populations of insects in their fields rather than eradicating them all with chemicals. They even release "beneficial insects" (insects that eat crop-destroying insects) on the plants whenever a certain pest threatens their crop. Weeds are one of the biggest obstacles for organic cotton farmers. Crews of laborers knock them back with weed eaters or hand pick them, and this additional effort can increase the cost of organic cotton fiber by up to 15 percent.

Perhaps the greatest hurdle organic farmers face is not in successfully growing cotton but in finding markets for their premium fibers. The clothing business is extremely competitive and consumers around the world are reluctant to pay more than they have to for a garment. Right now, Europeans, particularly Germans, are the only consumer group that understands the value of organically grown fibers and is consistently willing to pay more for a product because it has been produced in the least harmful way possible.

A handful of farmers and manufacturers are demonstrating that cotton production doesn't have to come at the expense of human and environmental health. If we are seriously committed to preserving the long-term health of our farmlands and the communities that surround them, we should give our support to these leaders in the field who are forging new paths toward an even higher quality of life. But until we start to perceive the things we buy differently, taking into account the environmental costs of every phase of production as well as the final product, these farmers and manufacturers face an extremely steep, uphill battle.

by the National Cotton Council

2. Cotton, Health, and the Environment

The National Cotton Council, based in Memphis, Tennessee, represents producers, merchants, manufacturers, and other members of the cotton industry.

The United States produces an average of 17 million bales of cotton on about 12 to 15 million acres in 16 states. Producing 25 percent of the world's cotton, the United States is by far the largest cotton exporter in the world.

People around the world can be assured that U.S. cotton and cotton textiles are produced and processed using the most advanced, safest, and most environmentally sound systems in the world.

Cotton producers live close to the environment—in rural areas, in the middle of their cotton fields, orchards, and pastures. The manner in which they farm their land affects them and their families first and most directly.

They are also business people. They cannot afford to squander their resources. Soil erosion, excessive water consumption, and unnecessary chemicals are all very expensive.

Today's efficient farming techniques spare vast areas of land from the plow. The United States grew more than 34 million acres of cotton in 1934 and produced almost 19 million bales of cotton fiber. In 1994, almost 20 million bales were grown on only 14 million acres—a 60 percent reduction in the area used for growing cotton. This is land that can be used for wetlands, wildlife reserves, pastures, and timber. With the implementation of new management techniques, the next generation could see three times the amount of food and fiber produced on the same amount of land.

Cotton is a safe and renewable resource. In the United States, its production and processing are conducted within a regulated system that provides strong protection for the consumer and the environment. It's true that chemicals are used on the crops. They are necessary to prevent insects and disease from spoiling crops; to keep weeds from using up the light, nutrients, moisture, and space around crops; and to add nutrients to the soil in the form of fertilizers.

Remember, too, that chemicals for controlling pests are used by almost everyone. These include insecticides that control roaches in people's homes, disinfectants that prevent the spread of germs, and fungicides that make roses and shrubbery more beautiful.

2. Cotton, Health, and the Environment (Cont'd.)

In the United States, pesticides for cotton are regulated by the Environmental Protection Agency (EPA) on the basis that it is a food crop. (About two-thirds of the harvested crop is composed of cotton seed, much of which is used for animal feed. Cottonseed oil is also used in many foods.) A recent European study determined that conventionally grown U.S. cotton was sufficiently free of pesticides that, theoretically, it could be used as a foodstuff in Germany. The EPA approves chemicals for use in agriculture after extensive testing for effectiveness and safety. To gain EPA approval, each product is subjected to more than 120 separate tests over a period of 8 to 10 years. These tests cost $35 million to $50 million per compound. Products failing any one of the tests are not approved for use in the United States.

Growers have a strong interest in applying agricultural chemicals safely, efficiently, and as wisely as possible, since they depend on these tools for a living. There are strict, legally enforceable penalties for misuse. Chemicals are generally applied by certified applicators based on recommendations of trained crop advisors. Specific products are prescribed only if their use is needed, if they are safe for the location, and if they will be spread by someone who follows any restrictions on the label.

Cotton growers in the United States participate in several organized programs that emphasize environmental awareness. "Cotton Cares" provides growers with an environmental self-assessment of their own farming practices. "Careful by Nature" is an educational program stressing wise, safe, and efficient application of chemicals. The High Cotton Award program recognizes producers for using available technology with a concern for the environment. And the cotton industry is a new partner in the EPA's Pesticide Environmental Stewardship Program, which emphasizes wise, safe, and efficient use of chemical products.

Some farmers have chosen to grow cotton organically. About 50,000 bales (0.3 percent of the U.S. crop) were produced organically in 1993, following production practices that use no artificial chemicals. Many materials used in this system, such as sulfur, nicotine, pyrethrins, bacteria, and viruses, are, in fact, pesticides, but they are derived from naturally occurring toxic substances. Typically, organic farming yields less cotton and requires more information, hand labor, time, and management than conventional cotton farming. Organic farmers charge more for their cotton than conventional growers because they can grow less on their land.

2. Cotton, Health, and the Environment (Cont'd.)

It's important to remember that whether land is cleared of its native plants and animals for shopping centers, apartment complexes, subdivisions, or agriculture, a human-made environment is created. It is no longer part of the natural ecology. Of the many human-made environments, farming, including cotton farming, is one of the more beneficial examples of stewardship of the land. This is true whether cotton is produced organically or conventionally. Agriculture depends on the biological health of the soil-water-life-based system. Farmers manage their farms using proven practices. These practices conserve the land as a sustainable resource that farming and urban families rely on for food, fiber, drinking water, recreation, and economic prosperity.

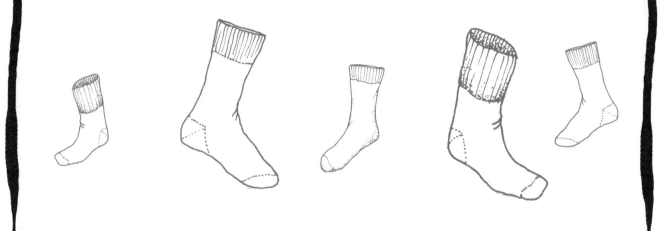

Hitchhiking on Hummers—Some species of small creatures called mites live inside flowers. To get from one flower to another, they "hitchhike" on hummingbirds. When a hummer sticks its long bill down into a flower to take a drink of nectar, the mites hop on board. They quickly scramble up the bird's bill and hide out inside its nostrils. When the hummer stops at another flower of the right kind, they scramble back down the bill and hop off into their new home.

by Carlos Vielma and Rachel Shutkin

3. Chatting About Cotton

The following is an e-mail exchange about organic cotton between two college students, Carlos Vielma and Rachel Shutkin.*

To: rachel@rvm.edu
From: carlos@zmu.edu
Date: October 23
Subject: The latest news

Hey, Rachel—
I know it's been a long time since I e-mailed you, but you know how busy college life gets. Classes and football are taking lots of time, and to make things crazier, I've started an environmental group at school. We spent last month trying to convince the cafeteria to carry organic food. Can you believe it—I, Carlos the junk food addict, am now eating almost 100 percent organic food. In my "Environment and Society" seminar, I learned about all the problems pesticides cause for wildlife and people. It grossed me out. So I figured I should put my food where my mouth is, and start eating the stuff I believe in.

Of course that doesn't just mean I eat organic food. It also means I'm trying to buy organic cotton clothing as much as possible. Did you know that something like 25 percent of the pesticides used in this country go to cotton crops? Organic cotton is kind of expensive, but at least you know you're not buying junk that's wrecking the planet. Now, if I could only find organic cotton football jerseys! That covers most of what I'm doing these days. How 'bout you? Zap me back sometime and fill me in.

Cheers,
Carlos

To: carlos@zmu.edu
From: rachel@rvm.edu
Date: October 25
Subject: The latest news—reply

Get out of here! My meat-and-potatoes, wide-receiver friend has turned into an eco-nut? I don't know how this happened to you, Carlos! You sound like you think the world is falling apart. But it's not. I agree that there may be some problems here and there but, overall, I think it's fine. I believe there are more important things to be thinking about than what your shirts are made of.

While you've been getting excited about saving the Earth, my friends and I have been trying to save people. Poor people. People who don't have the MONEY to afford fancy vegetables or high-priced clothes. Have you forgotten how hard it is for folks to make ends meet? It's hard to see why they should spend their hard-earned cash trying to save wildlife when they need it to save their own children and put food on their table.

I hope we can talk about this when you get back to town, Carlos.

Yours,
Rachel

*Carlos and Rachel are fictional characters.

by Yvon Chouinard, Patagonia

4. Food for Thought

Yvon Chouinard is president of Patagonia, an outdoor clothing retailer.
This essay appeared in the Patagonia Kids Catalog, spring/summer 1996.

During our company's ongoing environmental self-assessment we made an interesting discovery. The most damaging fabric we use to make Patagonia clothing, we learned, may be the one we consider the most "natural." Cotton.

Many parents prefer to dress their children in cotton because it's a comfortable, "100 percent pure" textile grown from a renewable resource. However, cotton can cause as much unnatural impact as do synthetics made from nonrenewable resources. It's the most pesticide-intensive of all crops. Heavy applications of chemicals for weed and insect control harm wildlife and pollute the air, groundwater, and soil. Where spraying occurs, farm workers and local residents experience serious health problems.

For several years farmers in the United States and other countries have been growing cotton without chemicals. Instead of relying on a never-ending feed of artificial fertilizers and toxins, organic cotton farmers build soil nutrients so the land itself sustains the crop. They reduce the risk of disease and pests through biological rather than chemical controls. As they gain experience, their cotton's quality and yields are becoming comparable to those of conventional growers.

As these cleaner, safer cotton farming methods emerge, we have a clear choice. All of Patagonia's cotton clothes will now be made from organic cotton.

Although farmers save money by not using chemicals, organic cotton requires more labor. That means that, for now, it's more expensive than conventional cotton. To keep prices down we've sacrificed some of our profit margin. We've dropped products that no longer make economic sense or fail to meet our standards for quality and performance. We're also relying on loyal customers to expand their definition of quality to include environmental responsibility. Otherwise someday, someone will pay the exorbitant bill for clean-up, soil restoration, and health care. That "someone" will be our kids.

The switch to organic cotton is in step with Patagonia's commitment to produce the finest clothing with the least environmental harm. It's a quiet revolution in consumer culture and a model for future generations: honoring the values that underlie sustainable agriculture with every clothing choice we make. We can support farmers and producers who truly put nature's well-being into this natural fiber.

by The Gap, Inc.

5. Picking Apart Cotton: Another Perspective

Gap, Inc. is a clothing retailer with more than 2,100 stores in the
United States, Canada, the United Kingdom, France, Germany, and Japan.

We know our activities have an impact on the environment. After all, our suppliers manufacture clothes. We use energy and natural resources. We make waste. But we also care about the Earth, and we're committed to minimizing our negative effects on the environment.

A lot of our products are made from cotton. Although it's a natural fiber, cotton growing is associated with many environmental problems, including pesticide use. In fact, more pesticides are used on cotton than many other crops. We're concerned about this. One of the many issues we're working on is how we can increase our use of pesticide-reduced cotton. But effecting change is not easy.

At Gap, Inc., we use the term "pesticide-reduced cotton" to include both certified organic and transitional cotton. Since 1993, Gap, Inc. has focused on three approaches to exploring pesticide-reduced cotton apparel: (1) developing partnerships with suppliers to use organic cotton in our garments, (2) sharing information with other apparel companies, and (3) researching pesticide-reduced cotton issues.

Between 1993 and 1995, three of our divisions—Gap, GapKids, and Banana Republic—purchased hundreds of thousands of pounds of certified organic and naturally colored cotton to make clothes, including T-shirts and polo shirts.

One of the biggest lessons we've learned from our experience, and that of other companies, is that price is a major barrier to wider use of pesticide-reduced cotton. Since 1993, certified cotton has cost between 15 percent and 50 percent more than conventional cotton. We have been told that reasons for the higher cost include: increased labor to inspect fields for insects; increased costs at the gin for special handling, such as cleaning the equipment to ensure that certified organic cotton isn't mixed with conventional cotton; the cost of the certification process; and the increased cost of capital (banks and insurance companies charge farmers more for loans and insurance because they perceive that organic farming may reduce yields as compared to conventional practices).

5. Picking Apart Cotton: Another Perspective (Cont'd.)

The higher cost of pesticide-reduced cotton fiber can result in higher prices for jeans, T-shirts, and other clothes made with that fiber. Customers who want organic cotton may end up paying more. Companies that offer pesticide-reduced cotton products may experience a drop in sales or profit margin because many customers are unwilling to spend more on these products, even if they know they have a better environmental record.

Another lesson we learned is that the supply of pesticide-reduced cotton is limited. In 1994, approximately 30,000 bales of certified organic cotton were produced worldwide. That's less than 0.1 percent of the total worldwide conventional production of cotton. Limited supply of pesticide-reduced cotton creates a dilemma: Cotton mills shy away from committing to buying it because of concerns that there won't be enough to meet their needs, and farmers grow less of it because there doesn't seem to be enough demand.

There is good news. Despite rumors of inferior quality, we discovered that there are no significant quality or technical performance differences between conventional and pesticide-reduced cotton. Customers can't see or feel the difference, either. From a business perspective, though, this isn't always a good thing: A customer who compares two apparently identical T-shirts may resist paying 20 percent more for the one with the organic cotton label.

The business and environmental issues surrounding pesticide-reduced cotton are complex. We've learned that there are no quick answers or easy solutions. We hope that our efforts will allow us to share information and ideas with other manufacturers and retailers, and to make pesticide-reduced cotton clothing accessible to more people.

by Patrick Leung, M.D.

6. Cotton Farming and Your Health

Dr. Patrick Leung is a pediatrician in Bakersfield, California, who specializes in the treatment of allergies. For his project studying the cotton defoliants, Dr. Leung's son Jason received first place awards in the Pharmacology/Toxicology Senior Division at the California State Science Fair, as well as the Navy/Marine Corps Distinguished Science Awards in 1995 and 1996. Jason is currently attending the University of Notre Dame.

In the late summer and early fall, many people living in the San Joaquin Valley in California experience annoying illnesses. The onset of symptoms—such as a running, stuffy nose, itchy eyes, and even difficulty in breathing—becomes a yearly ritual coinciding with the arrival of the harvest season. The San Joaquin Valley is one of the most productive agricultural regions in the world. Cotton is a leading commercial crop for the region. But with the sound and sight of small planes hovering over the farms each fall, it isn't surprising that many people blame their stuffy, asthmatic symptoms on the aerial spraying of agricultural chemicals.

During the growing of cotton, many varieties of chemicals, called *insecticides,* are used to control harmful insects. Another group of chemicals, called *defoliants,* are used on cotton plants to make the leaves drop.

In the old days, during the harvest, people were sent out to the fields to pick the cotton bolls by hand. It was laborious and back-breaking work. Nowadays, special machines, called cotton harvesters, are sent out to the field to collect the cotton bolls. But before the machines can start the harvesting, defoliants have to be applied to get rid of the leaves.

Farmers take special precautions when the defoliants are used in order to minimize any harmful effects on humans. Acute illness resulting directly from the cotton defoliants is rare. However, with the yearly recurrence of illnesses each fall, people have become very suspicious about whether the defoliants may be contributing to their itchy, stuffy symptoms. There is hardly any research done in this area to confirm or disprove their suspicion.

My son, Jason Leung, has been involved in science projects since fifth grade. When he was a high school sophomore, he became curious and wanted to know whether cotton defoliants caused *bronchospasm.*

158

6. Cotton Farming and Your Health (Cont'd.)

(Bronchospasm is the narrowing of the air passages in the lung that leads to difficulty in breathing in people suffering from asthma.) It would have been impractical to study the cotton defoliants directly on humans. After all, who would volunteer to breathe potentially harmful chemicals?!

So, instead, Jason did his study using earthworms! Earthworms have a simple neuromuscular system (that's the system through which nerves activate muscles). And it's similar to that of humans. Jason designed a simple apparatus to measure the muscular contractions of the earthworm. He reasoned that the earthworm's muscles would contract in response to the same things that would make humans' air passages constrict.

Over a three-year period, Jason tested different kinds of cotton defoliants. He was able to demonstrate that cotton defoliants could induce muscular contractions in the earthworm, probably by causing irritation of the sensory receptors (the nerve endings that detect outside stimulation). People with allergies and asthma have air passages that are more susceptible to irritation. So, Jason's study suggested that cotton defoliants might aggravate their symptoms by causing added irritation during the harvest season.

Other scientists at the California Environmental Protection Agency at Berkeley have made different findings. They analyzed the deaths of people living in the San Joaquin Valley from 1970 to 1990. In 1995, they published their findings, saying that although there was an increased death rate due to respiratory diseases, the increase was related more to the dust in the air than to the cotton defoliants.

As one can see, the issue is complicated. Some people suggest banning the chemicals in farming in favor of organic methods, such as using beneficial insects to control the population of insect pests. But that solution presents new, unanswered questions.

The success of farming is very important to ensure our food supply and strengthen our economy. However, the environment and the health of the general public should never be compromised. Every farm, whether using chemicals or organic means such as beneficial insects, should be managed responsibly. The proper risks and benefits should be carefully evaluated.

The Wonders of Worms—The world's 3,000 species of earthworms all help plants in the same ways. By loosening the soil with their tunnels, worms make it easier for plants to extend their roots, for oxygen to reach root tips, and for water to be held for plant use.

Directions: Fill out the "Sorting Out the Issues" chart, being sure to include every writer from the readings. Then answer the following questions using the back if you need more space.

 1. List several things that the different writers agree on regarding conventional and organic cotton growing.

 2. List several things that the writers disagree on regarding conventional and organic cotton farming.

 3. Which writers seem to have similar points of view?

 4. Which writers have the most different points of view?

 5. What are some of your thoughts about conventional and organic cotton growing now that you've read these essays?

 6. What other types of information would you need to form your own opinion about the conventional versus organic cotton farming dilemma? make a decision? Do you think you could find that information? What facts would you like to verify?

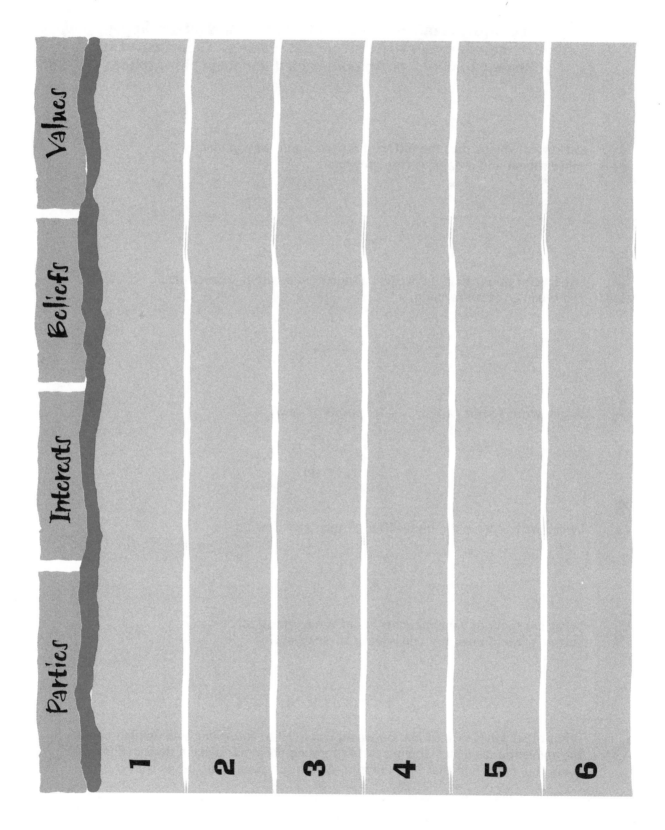

Values

Beliefs

Interests

Parties

1 2 3 4 5 6

One way to tell plants apart is by looking at their leaves.

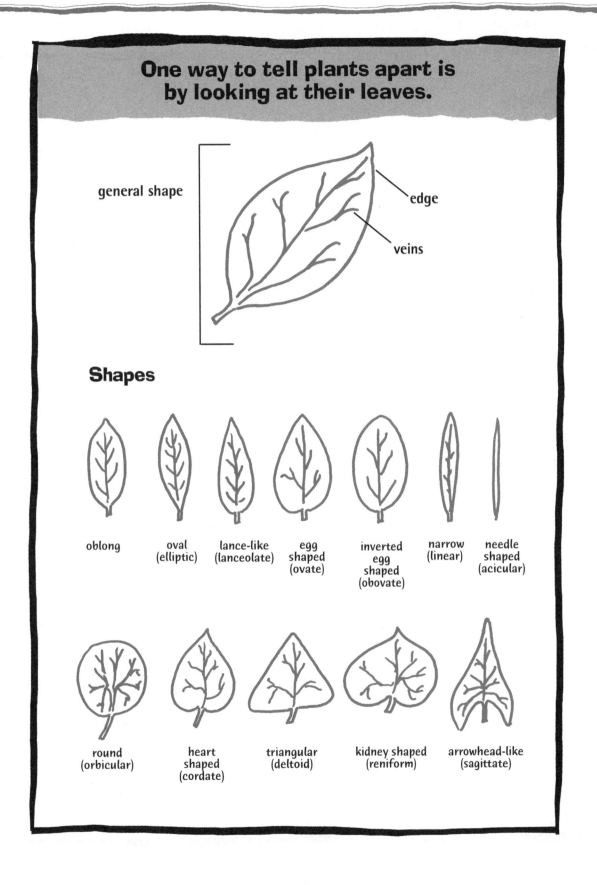

general shape

edge

veins

Shapes

oblong

oval (elliptic)

lance-like (lanceolate)

egg shaped (ovate)

inverted egg shaped (obovate)

narrow (linear)

needle shaped (acicular)

round (orbicular)

heart shaped (cordate)

triangular (deltoid)

kidney shaped (reniform)

arrowhead-like (sagittate)

Edges

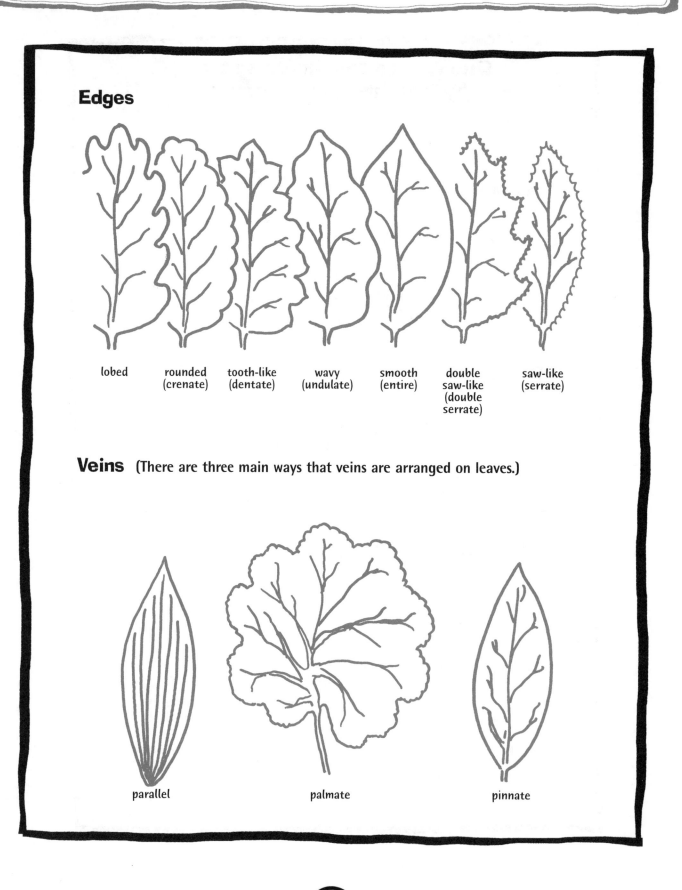

lobed   rounded (crenate)   tooth-like (dentate)   wavy (undulate)   smooth (entire)   double saw-like (double serrate)   saw-like (serrate)

Veins (There are three main ways that veins are arranged on leaves.)

parallel   palmate   pinnate

GRAPHING GREENS DATA LOG

PLOT NUMBER

SPECIES	1	2	3	4	5	6	7	8	9	10

DATA SUMMARY TABLE										
NEW SPECIES (first seen in sample area; ⊗s this plot)										
TOTAL NUMBER OF SPECIES (all ⊗s up to now)										
PLOT AREA (sq. yard)	1	1	1	1	4	4	4	16	16	16
TOTAL SAMPLE AREA (total of plot areas in sq. yards)	1	2	3	4	8	12	16	32	48	64

he island of Java is part of the island chain that makes up Indonesia. It is one of the world's most densely populated areas, with about 2,000 people packed into every square mile. (In the United States, there are about 70 people per square mile.) Scientists think the island was once completely covered with tropical forests, but because so many people have been living there, most of the forests have been destroyed. Today, the forests of Java can be found only in small fragments. A few reserves have been set aside to help preserve biodiversity on the island by protecting some of the remaining forest fragments. These reserves are like islands because they are surrounded by cities, farms, and cattle pastures, and they are isolated from other patches of forest. Scientists want to know how big these forest reserves need to be to support all the species native to the island.

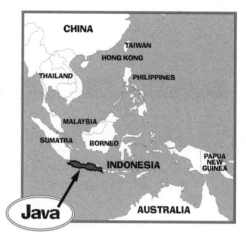

Scientists went to Java to study raptors that lived in the different forest reserves. Raptors are birds of prey that have excellent vision; strong legs and feet; sharp, curved claws for attacking their prey; and a hooked beak for tearing food into chunks. Java has many different species of raptors, including serpent eagles, honey buzzards, goshawks, sparrow hawks, black eagles, hawk eagles, and kestrels. These birds need more space than many other species so they can soar through the air to hunt for the insects, small rodents, snakes, dead animals, and other birds that make up their diet. Since they need such large areas, they can be in danger when their habitat becomes fragmented. The researchers spent a lot of time studying the raptors and how much space they needed. The graph below is one of the graphs they made to help organize the data they collected. Look at the graph, and then use it to answer the questions. (Use the raptors' perches as data points.)

- What is the title of the *x*-axis?

- What is the title of the *y*-axis?

- What is the average number of birds the researchers saw every hour in the 100 km² reserve? In the 150 km² reserve? In the 350 km² reserve? In the 500 km² reserve?

- In general, did the researchers see more birds in the large reserves or the small reserves?

- Based on this graph, and some of their other research, the scientists concluded that larger reserves are needed on the island. Why might they have thought that many small reserves on the island would not support the raptors?

- What title would you give the graph?

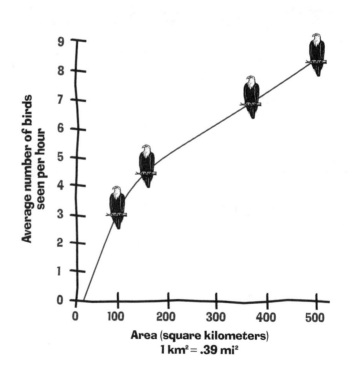

Windows on the Wild: Biodiversity Basics

World Wildlife Fund

Your group demand for chewing gum

Price	Number of Buyers
$0.10	
$0.20	
$0.30	
$0.40	
$0.50	
$0.60	
$0.70	
$0.80	
$0.90	
$1.00	
$1.10	
$1.20	
$1.30	
$1.40	
$1.50	
$1.60	
$1.70	
$1.80	
$1.90	
$2.00	

Worksheet 1

Everybody loves a bargain. That's why the lower the price of an item, the more likely it is that people will buy it. In the case of gum, more people will buy more gum at 10 cents a pack than at 50 cents a pack. If you fill in these numbers about how much people are willing to pay for gum, you'll get something called a *demand curve*. It shows that as the price of gum drops, more people are willing to buy a pack. The numbers provided below have been compiled from many different groups and show the total amount of gum people are willing to buy at certain prices.

Price and Number of Packs Demanded			
$2.00 = 0 packs	$1.50 = 5 packs	$1.00 = 10 packs	$0.50 = 15 packs
$1.90 = 1 pack	$1.40 = 6 packs	$0.90 = 11 packs	$0.40 = 16 packs
$1.80 = 2 packs	$1.30 = 7 packs	$0.80 = 12 packs	$0.30 = 17 packs
$1.70 = 3 packs	$1.20 = 8 packs	$0.70 = 13 packs	$0.20 = 18 packs
$1.60 = 4 packs	$1.10 = 9 packs	$0.60 = 14 packs	$0.10 = 19 packs

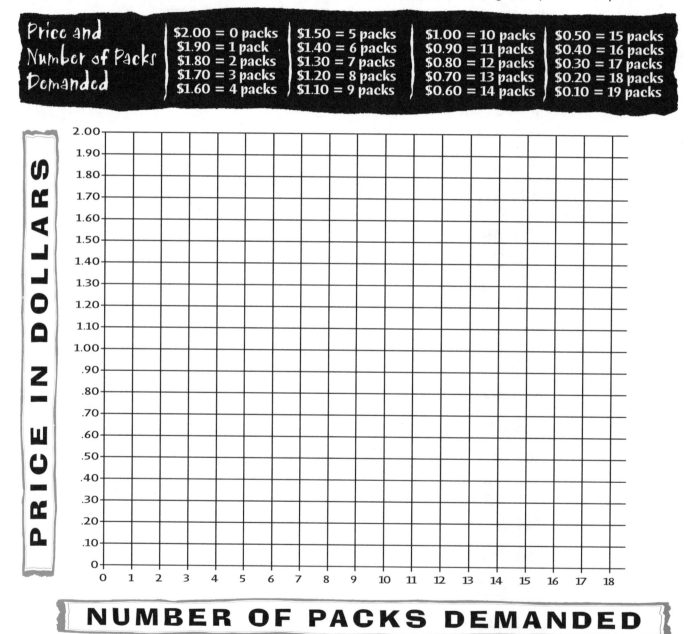

PRICE IN DOLLARS

NUMBER OF PACKS DEMANDED

Since people will buy more packs of gum at a cheaper price, why don't stores drop their prices to sell as much gum as possible? The reason is that store owners need to charge enough so that they recoup what it costs to purchase and resell the gum. Sometimes by selling a lot of gum, they can make up the costs. But there's a limit to how low they can set the price and still cover costs and make a profit. Think about what it takes to purchase and resell a pack of chewing gum. What do you think are some of the costs?

Worksheet 2

To cover their costs, gum producers can't sell their gum for less than a certain price. But if, because of demand, the price can go higher, producers may supply more, and new gum producers may be attracted into the business. This situation creates something called a *supply curve*. Below is a supply curve showing how many packs of gum firms will produce, depending on the price. Use it to answer the questions at the bottom of the page.

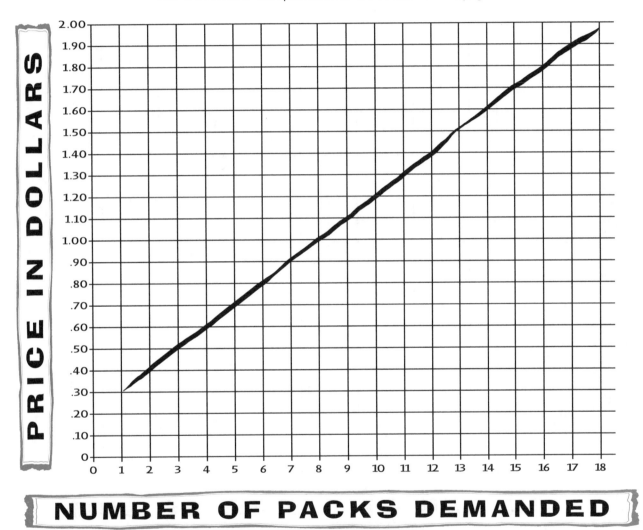

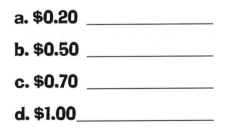

NUMBER OF PACKS DEMANDED

How many packs of gum will gum companies produce at:

a. $0.20 _____

b. $0.50 _____

c. $0.70 _____

d. $1.00 _____

Worksheet 3

Draw both the **supply curve** and the **demand curve** from Worksheets 1 and 2 on the graph below. The place where they meet is the **price**. The price is determined by all the interactions between potential buyers (demand) and potential sellers (supply) in the gum market. And the market (including wholesale and retail) is not one place, store, or company—it is represented by all the stores and companies that sell gum and all the customers who buy the gum.

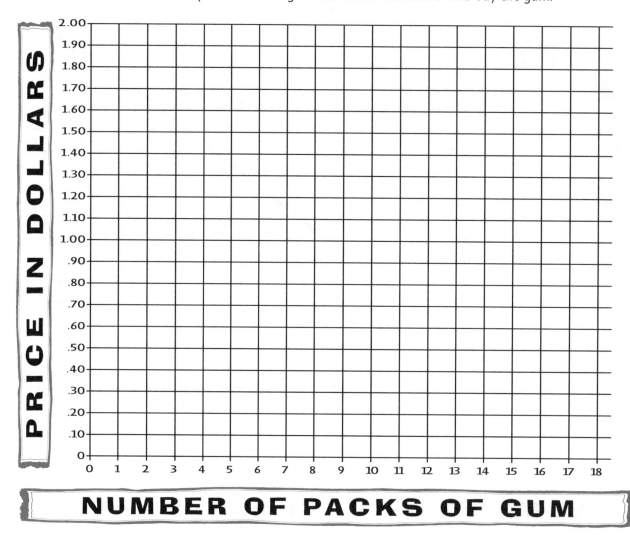

PRICE IN DOLLARS

NUMBER OF PACKS OF GUM

According to your data, what is the price of a pack of gum?_____

Is the price more or less than the price in real life? Why?_____

_____

_____

STUCK ON GUM WORKSHEET

Part 1

What's the connection between biodiversity and buying a pack of gum? You'll start to find out as you learn the story of how gum was produced in the past and how it is produced today. After you read or listen to Part 1 of the story of gum, answer the questions below.

1. At one time a large amount of gum, or chicle, was gathered in the rain forests of Central America. On the table below, name three positive and three negative effects this harvest might have had on the people and the environment in the area where it was gathered.

POSITIVE EFFECTS	NEGATIVE EFFECTS
1	1
2	2
3	3

2. When chicle was replaced by synthetic materials, what might have happened to the people who had gathered chicle? How could the change have affected the culture of the communities that relied on the sale of chicle?

Part 2: Breaking It Down

1. **The drawing below shows an average stick of gum. Write which ingredient each part represents.**

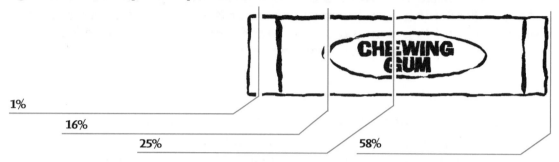

1%

16%

25%

58%

2. **Today gum is made from a variety of sugars, plant oils, and other chemicals. Name some of the costs involved in making gum.**

3. **Name some of the positive and negative effects that today's gum production can have on people and the environment. What other information do you need to answer this?**

POSITIVE EFFECTS	NEGATIVE EFFECTS
1	1
2	2
3	3

4. **Although gum base is no longer made from chicle, most of the other things that make up a stick of gum are made from products that come from plants. On the back, list five plants that add flavor or sweetness to a stick of gum.**

TACKLING TRADE-OFFS

Every product you buy is connected to the environment in some way. All products either use natural resources as a raw ingredient or cause some type of pollution as they are produced, used, or thrown away. People around the world—from business owners to consumer groups—are looking at how to make products that have fewer negative impacts on the environment. At the same time they're looking at the role consumers play in shaping what people buy and sell.

The following questions will get you thinking about some of these trade-offs:

1. **If producers stopped making gum, name some of the positive and negative effects this might have on people, the environment, and the economy.**

2. **Think about the products you and your family buy every week. What are several ways your buying habits affect people and environments around the world? How could you find out more about the impact of your consumer choices?**

3. **What types of careers are involved in making sure that producers, consumers, and the environment are protected in our society?**

INTRODUCING THE FACTORS OF PRODUCTION

What does it take to make something?

Natural Resources	**Labor**	**Capital**	**Management**
The raw materials needed to produce a product (e.g., timber, minerals, water)	The human effort that goes into producing a product, including muscle power and brain power	The equipment and structures and skills needed to turn raw materials and labor into finished products	The process that organizes and manages the labor, raw materials, and capital to produce and sell finished products

Look at the list of activities in the boxes below, and decide which of the four factors of production each activity matches best.

Production Factor 1
Julie, a local high school student, runs the cash register at the corner store.

Production Factor 2
Gordon, a citizen of Jamaica, comes to the United States every year to harvest ripe sugar cane and earn extra money for his family.

Production Factor 3
Celeste has been a beekeeper since she was a little girl. She looks after about 20 hives that produce wax and honey.

Production Factor 4
Giant metal mixers stir the sweetener, gum base, softener, and flavorings together.

Production Factor 5
Rafael is an artist. He designed the flashy new gum wrapper for the super-sized gum packs.

Production Factor 6
Peppermint grows wild in many parts of the state. It is also grown commercially.

Production Factor 7
Bob owns and operates a farm that grows corn.

Production Factor 8
John manages a manufacturing plant that makes gum base.

Production Factor 9
At the paper processing plant, timber is converted into thick sheets used to wrap paper products.

Production Factor 10
Selim drives a truck to transport essential plant oils to different factories where the oils go into soaps, perfumes, and food.

Production Factor 11
Powdered sugar coats each stick of gum and keeps it from sticking to others.

Production Factor 12
Computers automatically track the gum factory's production output, sales, and inventory.

Production Factor 13
Prairies have fertile soil that is perfect for growing corn and sugar beets.

Production Factor 14
A machete is used to cut slits in the side of the sapote tree to collect sap.

Cut out the pyramid and paste it on a large piece of construction paper. Then cut out all of the boxes on "Introducing the Factors of Production" and glue them around the pyramid near the appropriate level. Next, draw a line from each activity to where it fits on the pyramid. (Hint: Each activity could be matched to more than one factor of production.)

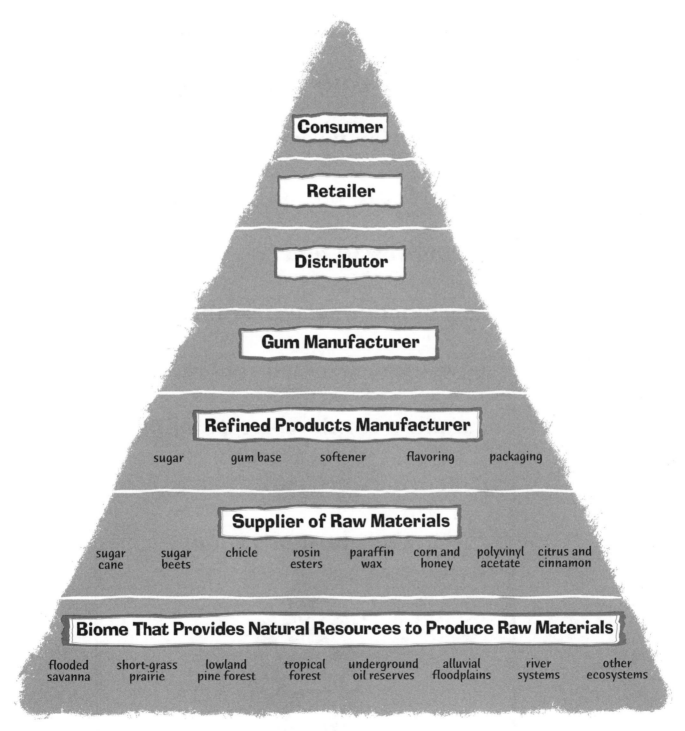

How Can We Pro

The activity pages in this section investigate how people are working to protect biodiversity and create more sustainable societies—through education, research, and futures studies. For the corresponding activities, see pages 322-357 in the Educator's Guide.

ct Biodiversity?

©David Wall/Lonely Planet Images

*"You really can change the world
if you care enough. . . . You have
an obligation to change it. You
just do it one step at a time."*

**–Marian Wright Edelman,
attorney, civil rights activist**

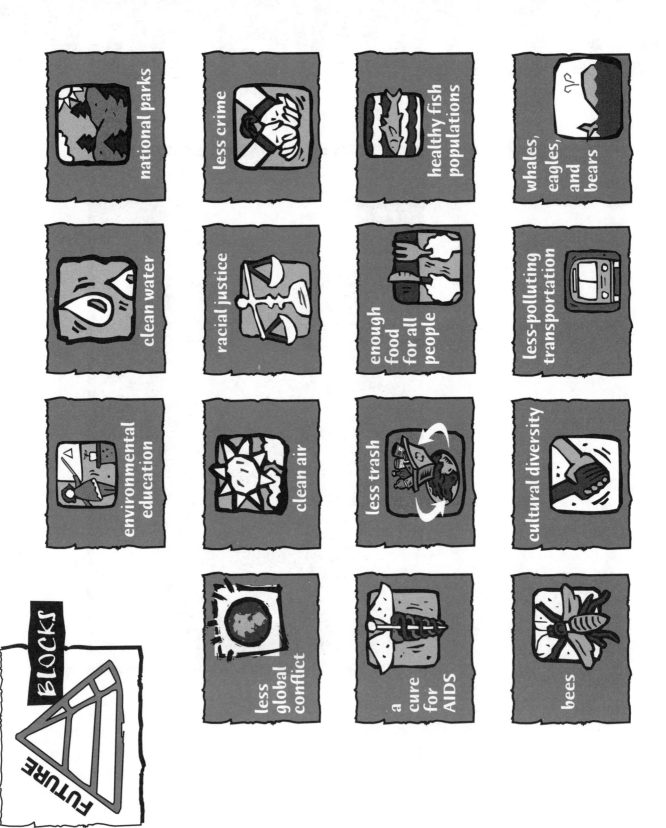

national parks

less crime

healthy fish populations

whales, eagles, and bears

clean water

racial justice

enough food for all people

less-polluting transportation

environmental education

clean air

less trash

cultural diversity

BLOCKS

FUTURE

less global conflict

a cure for AIDS

bees

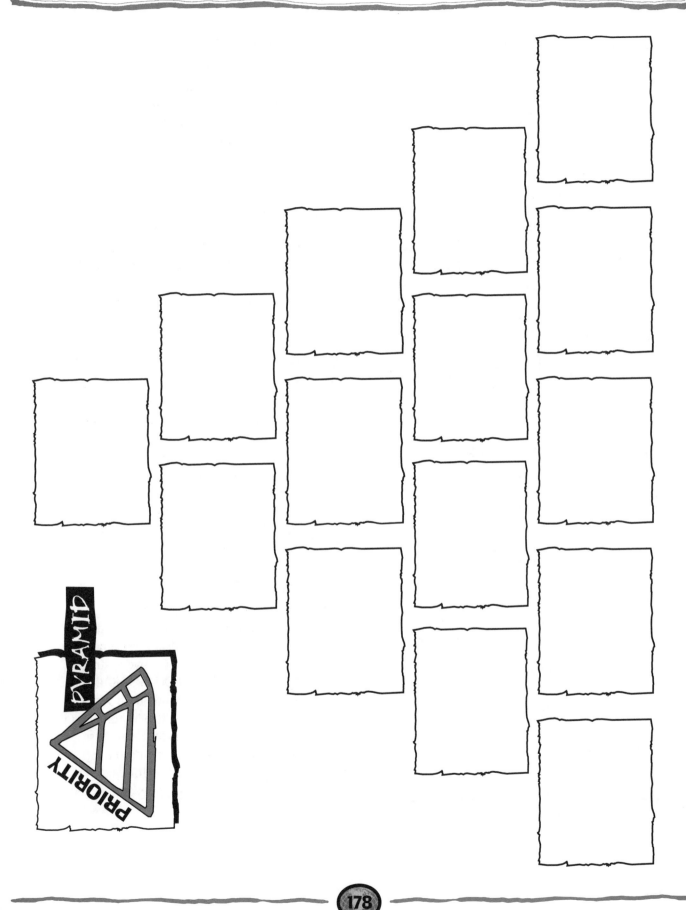

Bee Good to Your Lips

A company called Burt's Bees purchases beeswax for the production and sale of lip balm. That's good news for beekeepers across the United States who use the income from beeswax sales to better manage their honey bees.

Most people don't know it, but honey bee populations have declined in the last few years. If the trend continues, the cost to farmers could be billions of dollars a year. The reason? Bees are very important for pollinating crops. Saving bees will require changes on many fronts—from reducing pesticide use to saving bee habitats. And buying some lip balm from Burt's Bees can't hurt, either.

Now, That's More Lichen It!

Forest scientists need lots of high tech instruments to measure air quality, right? Well, traditionally, they *have* had to spend thousands of dollars a year to buy and operate electronic air-monitoring instruments. But now they have a new, more cost-effective instrument: lichens! Lichens are actually two organisms in one: a fungus and either an alga or a bacterium. Together these "partners," which grow on rocks and other surfaces, can live in some of the harshest environments on Earth—including the frigid reaches of Antarctica.

Tough as they are, though, lichens are very sensitive to air pollution. A botanist discovered how certain lichens respond to three different air pollutants—ozone, sulfur dioxide, and nitrogen oxide. By monitoring where lichens are growing and how healthy they are, scientists can draw conclusions about the presence of these pollutants in an area. And many forest scientists are doing just that!

No Time for Crime

Want to be a crime buster in your community? Maybe you should build a park! Studies show that parks and gardens have a big influence on community safety. After police helped residents of a Philadelphia neighborhood clean up vacant lots and plant gardens, crime in the precinct dropped by 90 percent.

Just giving people something to do can fight crime too. When basketball courts in Phoenix are kept open until 2:00 AM during the summer, crime goes down as much as 55 percent. And in Fort Myers, Florida, juvenile arrests decreased by nearly one-third after the city started an academics and recreation program for local students.

A New Way of Doing Business

The staff of Aveda Corporation is talking trash—how to reduce it, that is. They've decided to reduce the packaging of their shampoos, lotions, and other body care products. They're also using recycled paper, composting their cafeteria waste, and supplying reusable dishes to their employees. Together, these measures have helped the company cut solid waste per employee by 29 percent. And they've reduced solid waste per gallon of product by 66 percent.

City Sewer Savvy

Kids around the country are spray-painting messages on city streets. Are they breaking the law? No way! Working with local governments, they're helping to save their local rivers, streams, and bays. By painting "DON'T DUMP: DRAINS TO BAY" (or other waterway, depending on where they live) on city storm drains, they're reminding residents that pollutants dumped down these drains flow into local waterways. And those waterways are important habitats for wildlife—not to mention critical fishing spots and valued places for recreation.

Ending the Conflict

Throughout history and around the world, many wars have been fought over natural resources such as land, forests, and water. But now a number of groups are trying a new approach to ending conflicts before they escalate into fighting. This new approach is called conflict management. In conflict management, disagreeing groups get together with an impartial party to discuss their concerns. Each side is asked to listen closely to the other side. And the impartial party helps clarify exactly what each side seems to be asking for. In many cases, once groups get beyond their anger and frustration, they find that their needs can be met without further conflict. So far, conflict management has been used to address disputes in the Middle East over water resources, in the Amazon rain forest over forest use, and in many other regions worldwide.

The Everglades for Evermore

The future is looking brighter for alligators, ibises, panthers, and other creatures living in Florida's Everglades National Park. Ever since it was established in 1947, the park has suffered from increasing pollution, decreasing water flow, and other problems caused by surrounding farms and cities. As the years passed, people began to realize that merely setting aside land for a park isn't enough to protect the area and its wildlife. So engineers, biologists, political leaders, and community members decided to team up to design a huge ecosystem recovery plan for the Everglades. The plan calls for cleaning up pollution, restoring the natural water flow to large parts of the region, and converting large areas of farmland

back into marsh. If it's carried out, this largest-ever recovery project could be a model for restoring degraded habitats all over the world.

A Breath of Fresh Air

After incinerators and other waste treatment plants were built in the predominantly lower-income neighborhood of Chester, Pennsylvania, people started getting sick from the fumes. As local citizen Zulene Mayfield learned, toxic facilities are often built in minority and under-privileged communities. But Mayfield didn't think that she and her neighbors should have to breathe polluted air. So the concerned citizens formed the Chester Residents Group. The group got one company to pay $356,000 for pollution violations. They also helped pass a law so that no new polluting waste treatment facilities can be built in the area. And they convinced companies to re-route noisy trucks away from the neighborhood.

The Sorcerer's Apprentices

How do scientists discover how to make new medicines from tropical plants? One of the best ways is to talk to local shamans—or medicine men and women—who show them the healing plants in the region. Then drug researchers can test these plants in their labs to see if they are effective in treating various diseases.

But the knowledge and wisdom of these shamans is disappearing as the shamans grow old and die. So an organization called Conservation International has helped to develop a Shaman's Apprentice Program. The program encourages younger people within a tribe to learn about medicinal plants from the elder shamans. Apprentice programs are up and running in Costa Rica, Guyana, Suriname, and Brazil. The programs may keep local traditions alive and help drug companies find cures for some of the world's deadliest diseases.

Songbirds on the Net

A student in Michigan logs in on the Internet and enters her observation: yellow warbler spotted today! She's one of many students participating in an environmental education program called MISTNET (Migration Information Songbird Tracking NETwork). Participants track the migration of several songbird species between their winter habitats in the tropics and their summer habitats in the United States and Canada. The students are helping scientists keep an eye on songbird populations, which are decreasing because of habitat loss in both areas and along the migration corridors. These students are learning a lot about birds and migration. And they're participating in a project that gives a whole new meaning to the phrase "web of life."

Carving Out a Niche in the Desert

The Seri Indians of northwestern Mexico carve and sell incredible sculptures of local wildlife out of a valued local tree called ironwood. But for a while, Seri carving was threatened by industries that also used ironwood trees. That was bad news not only for the Seri, but also for local wildlife: Ironwood trees provide shade, shelter, and nutrients to hundreds of desert animals. So a group of scientists and policy makers from the United States and Mexico joined together to help the Seri. They toughened laws to crack down on other uses of the ironwood tree, focused public attention on Seri arts and crafts, and helped the Seri diversify their carving materials. They hope these measures will keep local culture—and nature—thriving for years to come.

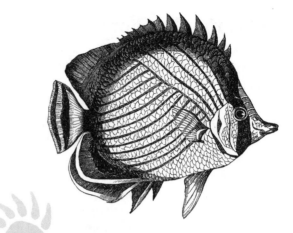

Reef Solution

Local fishers in villages in southern Thailand had a problem—a BIG problem. Large ships equipped with very efficient nets were scooping so many fish out of the sea to sell to people in other countries that the local fishers were having trouble catching enough to survive. So villagers joined together to form a group to analyze the situation and to look for solutions. The fishers met with members of the government and with owners of the big boats to explain their troubles, and put rules in place to keep anyone from catching too many fish. They also built artificial reefs near their villages by piling up rocks, cement, and wood offshore. These reefs created lots of nooks and crannies—great places for fish to live. Fish populations around the reefs have increased dramatically. And big fishing boats avoid the reefs because they don't want the reefs to wreck their nets.

A New Crop of Farmers

Fred Kirschenmann grows wheat, rye, oats, millet, and sweet clover on his North Dakota farm without using any chemical fertilizers or pesticides. He's part of a growing group of farmers in the United States practicing something called sustainable agriculture. These farmers want to farm in a way that is good for the environment and good for other people. As they point out, conventional farming may lead to bigger and bigger harvests of food, but it does so with methods that pollute the air and water, degrade the soil, hurt wildlife, and reduce diversity. In the long run, conventional farming may strip the land of its growing power. So these new farmers don't use pesticides or big machinery, and they grow a diversity of food crops to reduce the harmful effects that their activities might have on our environment.

Rediscovering the Bicycle

Bikes are a great solution to transportation problems. They run on human power, so they're completely nonpolluting. They're quiet. They don't need much space for parking or nearly as much road space as cars. But in most communities worldwide, biking can be inconvenient—even hazardous—if streets have been designed just for car use. That's why the city of Palo Alto, California, has built 40 miles of bike paths, including a 2-mile bikes-only boulevard in the center of town. The city requires new buildings over a certain size to build bike racks. And Xerox Corporation, located in Palo Alto, even offers towel service in its shower room for bikers. Maybe that's why 15 percent of Xerox's local employees cycle to work—one of the highest rates in the country!

Laws for Claws

It pulled the California sea otter, the peregrine falcon, and the red wolf back from the brink of extinction. It helped the nation's bald eagle population increase tenfold in just 25 years. What is this lifesaver? A piece of legislation called the Endangered Species Act (ESA). The ESA was passed in 1973 to protect all species in the United States from extinction. Using scientific information, federal agencies determine which species are "threatened" or "endangered." Then they come up with plans to help the species recover, say, by protecting habitat and banning hunting. So far, according to the U.S. Fish and Wildlife Service, populations of more than half of the 1,177 plants and animals the act protects in the United States are stable or growing—including black-footed ferrets, brown pelicans, and gray whales.

EASTER'S END

by Dr. Jared Diamond

Among the most riveting mysteries of human history are those of vanished civilizations. Everyone who has seen the abandoned buildings of the Khmer, the Maya, or the Anasazi is immediately moved to ask the same question: Why did the societies that erected those structures disappear?

Among all vanished civilizations, that of the former Polynesian society on Easter Island remains the most mysterious. The mystery stems especially from the island's gigantic stone statues and its impoverished landscape.

Easter Island, with an area of only 64 square miles, is the world's most isolated scrap of habitable land. It lies in the Pacific Ocean more than 2,000 miles west of the nearest continent (South America), and 1,400 miles from even the nearest habitable island (Pitcairn). Its subtropical location and latitude help give it a mild climate, while its volcanic origins make its soil fertile. In theory, this should have made Easter Island a miniature paradise, remote from the problems of the rest of the world.

The island derives its name from its "discovery" by the Dutch explorer Jacob Roggeveen on Easter (April 5) in 1722. Roggeveen's first impression was not of a paradise but of a wasteland.

The island Roggeveen saw was a grassland without a single tree or bush over ten feet high. Modern botanists have identified only 47 species of plants native to Easter, most of them grasses, sedges, and ferns. The list includes just two species of small trees and two of woody shrubs. With such plant life, the Islanders Roggeveen encountered had no source of real firewood to warm themselves during Easter's cool, wet, windy winters. Their native animals included nothing larger than insects, not even a single species of native bat, land bird, land snail, or lizard. For domestic animals, they had only chickens.

European visitors throughout the eighteenth and early nineteenth centuries estimated Easter Island's human population at about 2,000. As Captain James Cook realized during his brief visit in 1774, the Islanders were Polynesians. Yet despite the Polynesians' well-deserved fame as a great seafaring people, the Easter Islanders who came out to Roggeveen's and Cook's ships did so by swimming or paddling canoes that Roggeveen described as "bad and frail." The canoes, only 10 feet long, held at most two people, and only three or four canoes were observed on the entire island.

With such flimsy boats, the Islanders could not travel far offshore to fish. The Islanders Roggeveen met were totally isolated, unaware that other people existed. Yet the people living on Easter claimed memories of visiting the uninhabited Sala y Gomez reef 260 miles away, far beyond the reach of the leaky canoes seen by Roggeveen. How did the Islander's ancestors reach that reef from Easter, or reach Easter from anywhere else?

Easter Island's most famous feature is its huge stone statues, more than 200 of which once stood

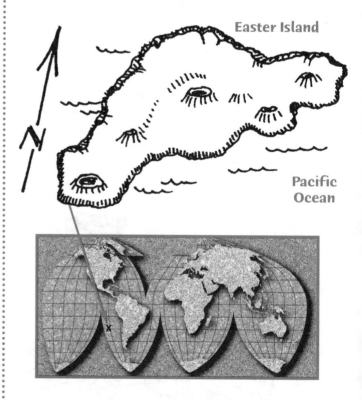

Easter Island

Pacific Ocean

on massive stone platforms lining the coast. At least 700 more, in all stages of completion, were abandoned in quarries or on ancient roads between the quarries and the coast, as if the carvers and moving crews had thrown down their tools and walked off the job. Most of the erected statues were carved in a single quarry and then somehow transported as far as six miles—despite their being as tall as 33 feet and as heavy as 82 tons. The abandoned statues, meanwhile, were as much as 65 feet tall and weighed up to 270 tons. The stone platforms were equally gigantic: up to 500 feet long and 10 feet high, with facing slabs weighing up to 10 tons.

Roggeveen himself quickly recognized the problem the statues posed: "The stone images at first caused us to be struck with astonishment," he wrote, "because we could not comprehend how it was possible that these people, who are devoid of heavy thick timber for making any machines, as well as strong ropes, nevertheless were able to erect such images." Roggeveen might have added that the Islanders had no wheels, no draft animals, and no source of power except for their own muscles. How did they transport the giant statues for miles, even before erecting them? To deepen the mystery, the statues were still standing in 1770, but by 1864 all of them had been pulled down, by the Islanders themselves. Why then did they carve them in the first place? And why did they stop? The statues imply a society very different from the one Roggeveen saw in 1722. Their sheer number and size suggest a population much larger than 2,000 people. What became of everyone?

The fanciful theories of the past must give way to evidence gathered by hardworking scientists in three specialties: archaeology, pollen analysis, and paleontology.

Modern archaeological digs on Easter have continued since 1955. The earliest evidence of human activities are from around A.D. 400 to 700. The period of statue building peaked around 1200 to 1500, with few if any statues erected after that time. Densities of archaeological sites suggest a large population: An estimate of 7,000 people is widely quoted by archaeologists, but other estimates range up to 20,000, which does not seem impossible for an island of Easter's area and fertility.

Archaeologists have also enlisted surviving Islanders in experiments aimed at figuring out how the statues might have been carved and erected. Twenty people, using only stone chisels, could have carved even the largest completed statue within a year. Given enough timber and fiber for making ropes, teams of a few hundred people could have loaded the statues onto wooden sleds, dragged them over lubricated wooden tracks or rollers, and used logs as levers to maneuver them into a standing position. Rope could have been made from the fiber of a small native tree called the hauhau. However, that tree is now extremely scarce on Easter, and hauling one statue would have required hundreds of yards of rope. Did Easter's now barren landscape once support the necessary trees?

That question can be answered by the technique of pollen analysis, which involves boring out a column of sediment from a swamp or a pond, with the most recent deposits at the top and relatively more ancient deposits at the bottom. The absolute age of each layer can be dated by radiocarbon methods. Then begins the hard work: examining tens of thousands of pollen grains under a microscope, counting them, and identifying the plant species that produced each one by comparing the grains with modern pollen from known plant species. For Easter Island, the bleary-eyed scientists who performed that task were John Flenley, now at Massey University in New Zealand, and Sarah King of the University of Hull in England.

Flenley and King's heroic efforts were rewarded by the striking new picture that emerged of Easter's prehistoric landscape. For at least 30,000 years before human arrival and during the early years of Polynesian settlement, Easter was not a wasteland at all. Instead, it was a subtropical forest of trees and woody shrubs, herbs, ferns, and grasses. In the forest grew tree daisies, the rope-yielding hauhau tree, and the toromiro tree, which furnishes a dense, mesquite-like firewood. The most common tree in the forest was a species of palm now absent on Easter but formerly so abundant that the bottom strata of the sediment column were packed with its pollen. The Easter Island palm was closely related to the still-surviving Chilean wine palm, which grows up to 82 feet tall and 6 feet in diameter. The tall, unbranched trunks of the Easter Island palm would have been ideal for transporting and erecting statues, and constructing large canoes. The palm would also have been a valuable food source, since its Chilean relative yields edible nuts as well as sap from which Chileans make sugar, syrup, honey, and wine.

What did the first settlers of Easter Island eat when they were not glutting themselves on the local equivalent of maple syrup? Recent excavations by David Steadman, of the New York State Museum at Albany, have yielded a picture of Easter's original animal world as surprising as Flenley and King's picture of its plant world. Steadman's expectations for Easter were conditioned by his experiences elsewhere in Polynesia, where fish are overwhelmingly the main food at archaeological sites, typically accounting for more than 90 percent of the bones in ancient Polynesian garbage heaps. Easter, though, is too cool for the coral reefs beloved by fish, and its cliff-girded coastline permits shallow-water fishing in only a few places. Less than a quarter of the bones in its early garbage heaps (from the period 900 to 1300) belonged to fish: Instead, nearly one-third of all bones came from porpoises.

On Easter, porpoises would have been the largest animal available—other than humans. Porpoises generally live out at sea, so they could not have been hunted by line fishing or spear fishing from shore. Instead, they must have been harpooned far offshore, in big seaworthy canoes built from the extinct palm tree.

In addition to eating porpoise meat, Steadman found, the early Polynesian settlers were feasting on seabirds. For those birds, Easter's remoteness and lack of predators made it an ideal haven as a breeding site, at least until humans arrived. Among the large numbers of seabirds that bred on Easter were albatross, boobies, frigate birds, fulmars, petrels, prions, shearwaters, storm petrels, terns, and tropical birds. With at least 25 nesting species, Easter was the richest seabird breeding site in Polynesia and probably the whole Pacific.

Land birds as well went into early Easter Island cooking pots. Steadman identified bones of at least six species, including barn owls, herons, parrots, and rails. Bird stew would have been seasoned with meat from large numbers of rats, which the Polynesian colonists accidentally brought with them; Easter Island is the sole known Polynesian island where rat bones outnumber fish bones at archaeological sites.

Porpoises, seabirds, land birds, and rats did not complete the list of meat formerly available on Easter. A few bones hint at the possibility of breeding seal colonies as well. All these delicacies were cooked in ovens fired by wood from the island's forest.

Such evidence lets us imagine the island onto which Easter's first Polynesian colonists stepped ashore some 1,600 years ago, after a long canoe voyage from eastern Polynesia. They found themselves in a pristine paradise. What then happened to it? The pollen grains and the bones yield a grim answer.

Pollen records show that destruction of Easter's forests was well underway by the year 800, just a few centuries after the start of human settlement. Then charcoal from wood fires came to fill the sediment cores, while pollen of palms and other trees and woody shrubs decreased or disappeared, and pollen of the grasses that replaced the forest became more abundant. Not long after 1400, the palm finally became extinct, not only as a result of being chopped down but also because the rats prevented its germination: Of the dozens of preserved palm nuts discovered in caves on Easter, all had been chewed by rats and could no longer germinate. While the hauhau tree did not become extinct in Polynesian times, its numbers declined drastically until there weren't enough left with which to make ropes. By 1955 only a single, nearly dead toromiro tree remained on the island, and even that lone survivor has now disappeared.

The 15th century marked the end not only for Easter's palm but for the forest itself. Its doom had been approaching as people cleared land to plant gardens; as they felled trees to build canoes, to transport and erect statues, and to burn; as rats devoured seeds; and probably as the native birds died out that had pollinated the tree's flowers and dispersed their fruit. The overall picture is among the most extreme examples of forest destruction anywhere in the world: the whole forest gone, and most of its tree species extinct.

The destruction of the island's animals was as extreme as that of the forest; without exception, every species of native land bird became extinct. Even shellfish were overexploited, until people had to settle for small sea snails instead of larger cowries. Porpoise bones disappeared abruptly from garbage heaps around 1500; no one could harpoon porpoises anymore, since the trees used for constructing large canoes no longer existed. The colonies of more than half of the seabird species breeding on Easter or on its offshore islets were wiped out.

In place of these meat supplies, the Easter Islanders intensified their production of chickens, which had been only an occasional food item. They also turned to the largest remaining meat source available: humans, whose bones became common in late Easter Island garbage heaps. Oral traditions of the Islanders are rife with cannibalism; the biggest insult that could be said to an enemy was: "The flesh of your mother sticks between my teeth." With no wood available to cook these new food sources, the Islanders resorted to sugar cane scraps, grass, and sedges to fuel their fires.

All these strands of evidence can be wound into a complete story of a society's decline and fall. The first Polynesian colonists found themselves on an island with fertile soil, abundant food, bountiful building materials, and all that they needed for comfortable living. They prospered and multiplied.

Eventually Easter's growing population was cutting the forest more rapidly than the forest was able to grow back. As the forest disappeared, the Islanders ran out of timber and rope to transport and erect their statues. Life became more uncomfortable—springs and streams dried up, and wood was no longer available for fires.

People found it harder to fill their stomachs as land birds, large sea snails, and many seabirds disappeared. Because timber for building canoes vanished, fish catches declined and porpoises disappeared from the table. Crop yield also declined, since deforestation allowed the soil to be eroded by rain and wind, dried by the sun, and its nutrients to be leached from it. Intensified chicken production and cannibalism replaced only parts of those lost foods. Preserved statuettes with sunken cheeks and visible ribs suggest that people were starving.

Surviving Islanders described to early European visitors how local chaos replaced organized government and a warrior class took over from the chiefs. By around 1700, the population began to crash toward between one-quarter and one-tenth of its former number. People took to living in caves for protection against their enemies. Around 1700 rival clans started to topple each other's statues, breaking their heads off. By 1864 the last statue had been thrown down and destroyed.

As we try to image the decline of Easter's civilization, we ask ourselves, "Why didn't they look around, realize what they were doing, and stop before it was too late? What were they thinking when they cut down the last palm tree?"

I suspect, though, that the disaster happened not with a bang but with a whimper. After all, there are those hundreds of abandoned statues to consider. The forest the Islanders depended on for rollers and rope didn't simply disappear one day—it vanished slowly, over decades. Perhaps war interrupted the moving teams; perhaps by the time the carvers had finished their work, the last rope snapped. In the meantime, any Islander who tried to warn about the dangers of deforestation would have been overridden by carvers, bureaucrats, and chiefs, whose jobs depended on continued deforestation. The changes in forest cover from year to year would have been hard to detect: Yes, this year we cleared those woods over there, but trees started to grow back again on this abandoned garden site here. Only older people, recollecting their childhoods decades earlier, could have recognized a difference. Their children could no more have understood their parents' tales than my eight-year-old son today can understand my wife's and my tales of what Los Angeles was like 30 years ago.

Gradually trees became fewer, smaller, and less important. By the time the last fruit-bearing adult palm was cut, palms had long since ceased to be of importance. That left only smaller and smaller palm saplings to clear each year, along with other bushes and treelets. No one would have noticed the felling of the last small palm.

It would be easy to close our eyes or give up in despair. If mere thousands of Easter Islanders with only stone tools and their own muscle power managed to destroy their society, how can billions of people with metal tools and machine power do better? But there is one crucial difference. The Easter Islanders had no books and no histories of other doomed societies. Unlike the Easter Islanders, we have histories of the past—information can save us. My main hope for my son's generation is that we may now choose to learn from the fates of societies like that of Easter Island.

Jared Diamond. Copyright © 1995. Reprinted with permission of *Discover.*

EASTER'S END

In 1722, a Dutch explorer named Jacob Roggeveen landed on Easter Island. Instead of a paradise, he found a wasteland.

But the most amazing thing the explorer found were huge stone statues lining the shore.

For more than 200 years, no one could solve the mystery.

Easter Island lies far out in the Pacific Ocean. It has an area of about 64 square miles, almost three times the size of Manhattan.

In the 1700s, people thought explorers would find the island a tropical paradise—with lush forests and lots of wildlife.

Instead of busy cities, he found only 2,000 people—not a big population for an island paradise. And he noticed that the only domestic animals were chickens.

How did these enormous statues get there with no machines or ropes or horses?

There were no tall trees. And the only native wildlife he saw were insects and other tiny creatures. There was not a single bat, snail, songbird, or lizard.

These statues were more than 30 feet tall and weighed as much as 82 tons—that's five times taller than an average person and heavier than 40 elephants!

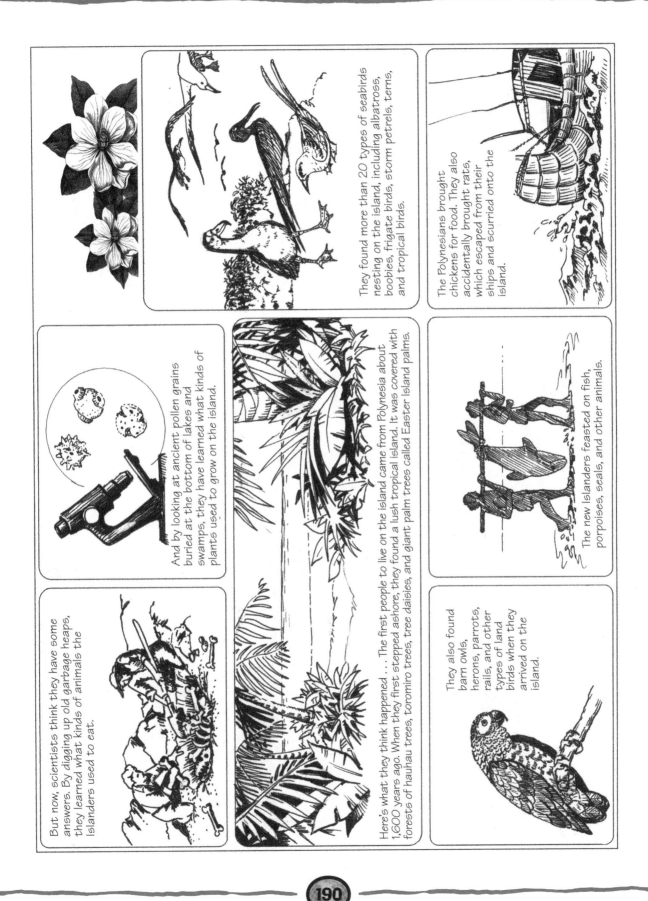

They found more than 20 types of seabirds nesting on the island, including albatross, boobies, frigate birds, storm petrels, terns, and tropical birds.

The Polynesians brought chickens for food. They also accidentally brought rats, which escaped from their ships and scurried onto the island.

And by looking at ancient pollen grains buried at the bottom of lakes and swamps, they have learned what kinds of plants used to grow on the island.

The new Islanders feasted on fish, porpoises, seals, and other animals.

But now, scientists think they have some answers. By digging up old garbage heaps, they learned what kinds of animals the Islanders used to eat.

Here's what they think happened . . . The first people to live on the island came from Polynesia about 1,600 years ago. When they first stepped ashore, they found a lush tropical island. It was covered with forests of hauhau trees, toromiro trees, tree daisies, and giant palm trees called Easter Island palms.

They also found barn owls, herons, parrots, rails, and other types of land birds when they arrived on the island.

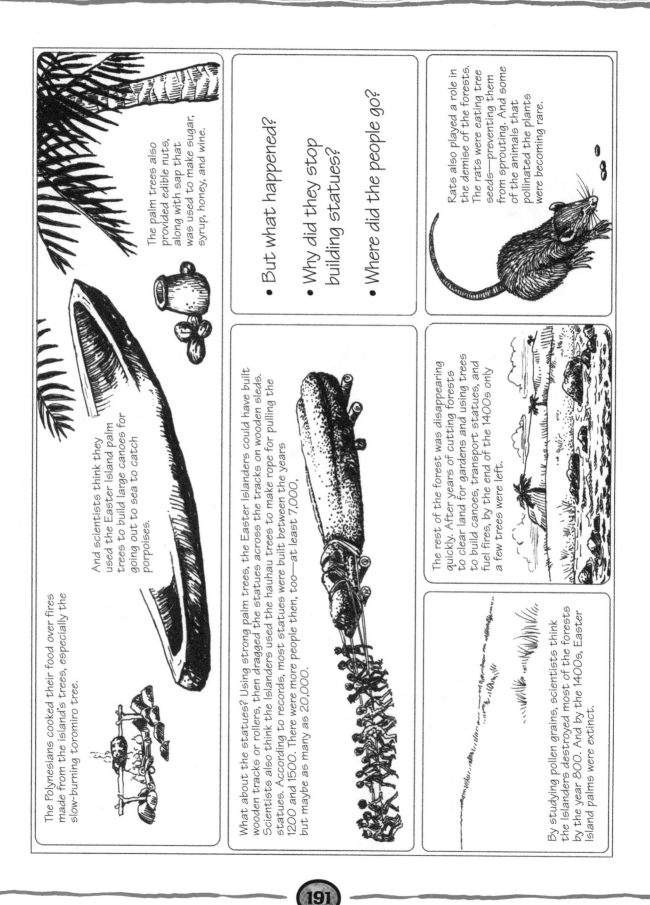

The Polynesians cooked their food over fires made from the island's trees, especially the slow-burning toromiro tree.

And scientists think they used the Easter Island palm trees to build large canoes for going out to sea to catch porpoises.

The palm trees also provided edible nuts, along with sap that was used to make sugar, syrup, honey, and wine.

- But what happened?

- Why did they stop building statues?

- Where did the people go?

Rats also played a role in the demise of the forests. The rats were eating tree seeds—preventing them from sprouting. And some of the animals that pollinated the plants were becoming rare.

What about the statues? Using strong palm trees, the Easter Islanders could have built wooden tracks or rollers, then dragged the statues across the tracks on wooden sleds. Scientists also think the Islanders used the hauhau trees to make rope for pulling the statues. According to records, most statues were built between the years 1200 and 1500. There were more people then, too—at least 7,000, but maybe as many as 20,000.

The rest of the forest was disappearing quickly. After years of cutting forests to clear land for gardens and using trees to build canoes, transport statues, and fuel fires, by the end of the 1400s only a few trees were left.

By studying pollen grains, scientists think the Islanders destroyed most of the forests by the year 800. And by the 1400s, Easter Island palms were extinct.

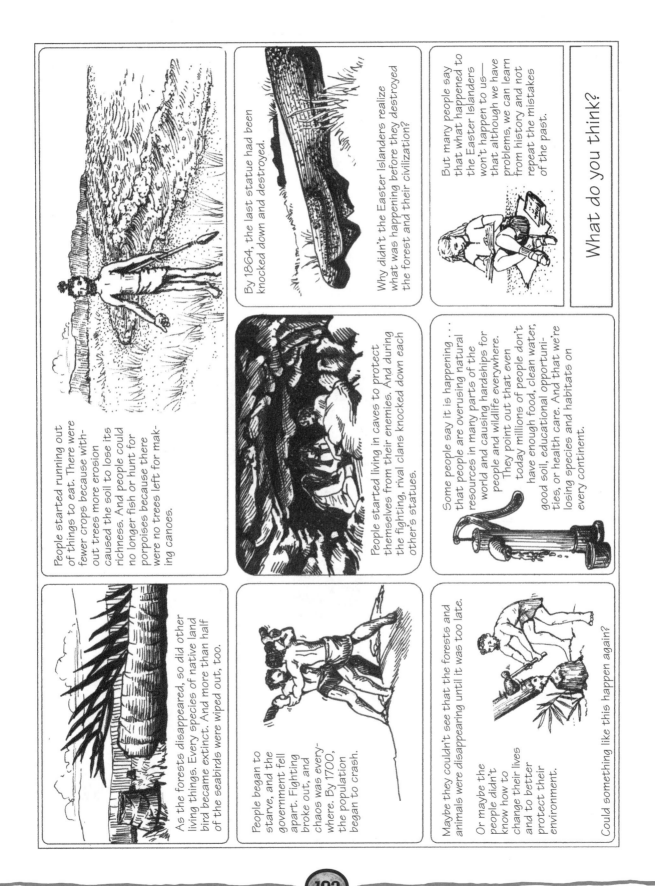

People started running out of things to eat. There were fewer crops because without trees more erosion caused the soil to lose its richness. And people could no longer fish or hunt for porpoises because there were no trees left for making canoes.

By 1864, the last statue had been knocked down and destroyed.

Why didn't the Easter Islanders realize what was happening before they destroyed the forest and their civilization?

But many people say that what happened to the Easter Islanders won't happen to us—that although we have problems, we can learn from history and not repeat the mistakes of the past.

What do you think?

People started living in caves to protect themselves from their enemies. And during the fighting, rival clans knocked down each other's statues.

Some people say it is happening . . . . that people are overusing natural resources in many parts of the world and causing hardships for people and wildlife everywhere. They point out that even today millions of people don't have enough food, clean water, good soil, educational opportunities, or health care. And that we're losing species and habitats on every continent.

As the forests disappeared, so did other living things. Every species of native land bird became extinct. And more than half of the seabirds were wiped out, too.

People began to starve, and the government fell apart. Fighting broke out, and chaos was everywhere. By 1700, the population began to crash.

Maybe they couldn't see that the forests and animals were disappearing until it was too late.

Or maybe the people didn't know how to change their lives and to better protect their environment.

Could something like this happen again?

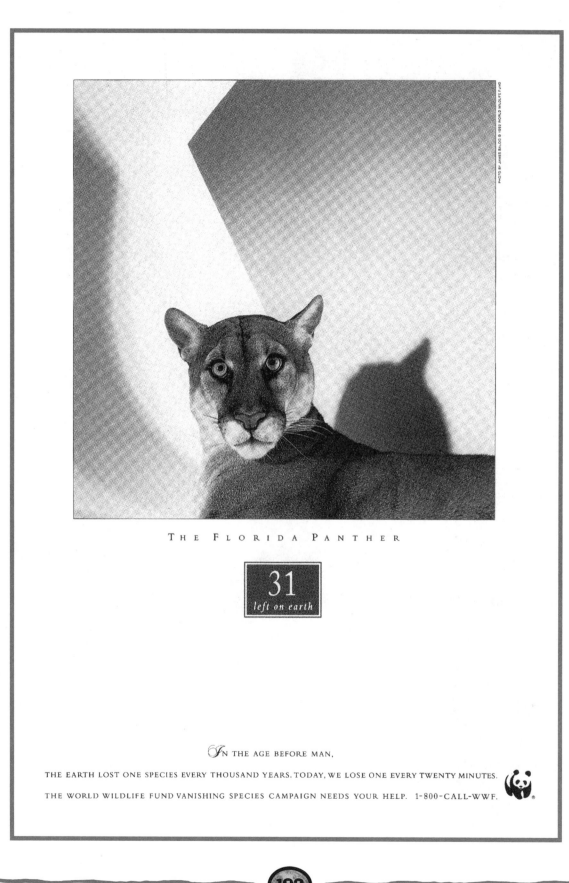

THE FLORIDA PANTHER

31
left on earth

IN THE AGE BEFORE MAN,

THE EARTH LOST ONE SPECIES EVERY THOUSAND YEARS. TODAY, WE LOSE ONE EVERY TWENTY MINUTES.

THE WORLD WILDLIFE FUND VANISHING SPECIES CAMPAIGN NEEDS YOUR HELP. 1-800-CALL-WWF.

All those in favor of

reducing deforestation,

raise your proboscis.

Save the

maggots.

(This isn't going to be easy, is it?)

Design by Huey/Paprocki, Ltd.

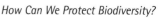

How this
heart-stopping
vampire bat
can help prevent
heart attacks.

Instead of giving you a heart attack, vampires may well help prevent it. The saliva from this South and Central American vampire bat was found to open clogged arteries twice as fast as conventional medicine. These compounds are now being developed to help prevent heart attacks.

Biodiversity
It's bigger than you think.

© 1999 Green Team Advertising, NY

He may seem
big and ugly
to you,
but to someone
with heart disease,
he's a prince.

The Houston Toad produces alkaloids, which can help prevent heart attacks. Now that's something worth kissing.

Biodiversity
It's bigger than you think.

© 1999 Green Team Advertising, NY

© 1999 Green Team Advertising, NY

What is Biodiversity and why should you care.

Biodiversity
It's bigger than you think.

Rosy Periwinkle
From the Madagascar Rosy Periwinkle comes Vinblastine and Vincristine. Since the introduction of these two drugs, childhood leukemia survival rates have increased 10% to 90%.

Sweet Berries
Move over sugar and saccarine! This Serendipity berry found in West Africa is 3000 times sweeter than sugar and has a lower calorie content.

One-Stop Supermarket
This Winged-Bean from New Guinea has been called the one-species supermarket. The entire plant is palatable, it produces a caffeine-free coffee-like drink, its beans have more protein than soy beans, it grows up to four meters in a few weeks and its nutrients help fertilize plants around it.

Wild Mexican Corn
A wild form of corn was discovered in the Cloud Forest of Jalisco, Mexico, which, when crossed with domestic corn, made it more resistant to disease. It not only helped feed millions, but is also contributing billions to the worldwide corn industry.

Armadillo Leper Healer
Cells from the spleen and liver of the armadillo were used to develop an anti-leprosy vaccine. This is one reason why leper colonies have almost been eliminated world-wide.

Vampire Heart Attack Preventer
What can give you a heart attack can now help prevent it. The saliva from this South and Central American Vampire Bat was found to open clogged arteries twice as fast as conventional medicine.

Hairy Tomato
What at first seemed like an insignificant discovery up in the highlands of Peru in 1962, this small "ugly and hairy" duckling of a tomato, proved to be swan in disguise. It has been found to be resistant to disease and has made the US tomato industry millions.

Do the Write Thing

Diane Jukofsky developed a love and respect for nature at an early age. She spent her summers on her grandparents' farm in the Ozark Mountains, where her grandmother taught her the names of wildflowers, birds, and stars.

Meanwhile, Diane's future husband, Chris Wille, was growing up in Oregon. Like Diane, he had a great love of wild things. By the age of six, he already had big collections of butterflies and seashells.

Now Diane and Chris work together as **environmental journalists** for the Rainforest Alliance, a nonprofit environmental group. Their job is to write articles about biodiversity and conservation issues in Latin America. The couple is convinced that people will be much more likely to join the effort to protect rain forests and other natural areas if they understand what biodiversity is, why it's important, and how quickly it is disappearing. As environmental journalists, Diane and Chris are doing what they can to help people learn about these issues.

The husband-and-wife team admits that they'll probably never get rich being journalists for a nonprofit environmental group. But their job has other rewards. As they put it, "The satisfaction you feel knowing you're making a difference is the greatest wealth."

Diane Jukofsky and Chris Wille

Lynda Richardson

Furniture for the Future

Furniture can be both beautiful and functional— but who ever heard of *smart* furniture? **Businessperson** Dani Sjahalam has. In fact, his company sells it. Dani is vice president of the Lynn-Nusantara Marketing Company in Oregon, a company that

Dani Sjahalam

sells furniture and other wood products in the United States, Canada, Mexico, and South America.

Dani became alarmed by the problems sometimes caused by harvesting certain species of trees in Indonesia and other tropical countries. Harvesting methods can be harmful to the forests where tropical trees grow, and the trees themselves are becoming rare in some areas because of overharvesting. So Dani contacted the Indonesian forestry department (Dani himself is Indonesian) and worked with them to identify forests where pine, teak, mahogany, and rosewood trees are being harvested in ways that will preserve the forests and trees for many future generations. He also worked with the environmental organization Rainforest Alliance, which agreed to call the wood harvested from these areas "smart woods." The products made from smart woods are labeled with a special Rainforest Alliance stamp to let consumers know that the wood is environmentally friendly.

Because his business now sells these smart woods, Dani feels that he's helping to educate buyers about the value of purchasing wood products that don't destroy tropical forests. And in the process, he's helping to protect biodiversity.

A Man with a Plan

Travel! Observe! Be creative! This is the advice Guillermo "Guillo" Rodriguez gives to students interested in doing what he does best—designing parks. Guillo, a **landscape architect** from Cuba, works in the Parks and Recreation Department in Durham, North Carolina. Creating new parks and sprucing up older ones is what his work is all about. And he strongly believes in building sustainable landscapes—areas that will remain beautiful for many years with little use of water, fertilizer, pesticides, or labor.

Guillermo Rodriguez

One way Guillo creates sustainable landscapes is by designing parks that require little maintenance and that can meet the needs of both people and wildlife. For example, he keeps lawn areas at a minimum to cut down on the need for mowing, watering, and fertilizing grass. In place of lawns he often plants low-maintenance wildflower meadows and hedgerows—areas that provide important habitat for wildlife. Guillo also works to preserve forested areas on park sites.

Guillo, who sees every project as "a new and fresh challenge," believes that people have the power to help design and improve the parks in their own communities. How about organizing regular park cleanup days? Or working with park authorities to design a butterfly garden? Who knows—maybe the landscape architect "bug" will bite you as hard as it bit Guillo!

Close Encounters

What would it be like to be face-to-face with a mountain gorilla in Rwanda? Or an Adelie penguin in Antarctica? How about a waxy tree frog in Venezuela? Gerry Ellis knows what it's like! Gerry, a **nature photographer** who lives in Portland, Oregon, has been snapping pictures of wild places and faces around the world for more than 15 years.

Gerry uses his camera to capture the wonder and diversity of life, and the amazing connections among different life forms. He believes that all species are tied together and to their environment in thousands of ways, forming interactions as diverse as our imaginations.

Although Gerry spends much of his time out in the wild with his camera, he points out that his office work is equally important. Preparing for a single shoot can mean repairing gear, researching the species and the conditions of the site where he'll be working, and arranging for transportation and lodging (a task that's not always easy in the far-off places where he travels). Insisting that there is "absolutely no substitute for learning from experience," Gerry suggests that those considering a biodiversity-related career do volunteer work. He also recommends getting out into nature—as often as possible.

Gerry Ellis

Mapping the Future

What do maps have to do with endangered species? Plenty, if you're a **mapping specialist** like Lata Iyer. Born and raised in India, she now works for an environmental group called Conservation

Lata Iyer

International in Washington, D.C. As a mapping specialist, Lata collects information about remaining forests, human population centers, and expanding agricultural uses of the land. Then she uses this information to make maps that show areas with habitats that should be protected. These maps are used by biologists, social scientists, government officials, and mapping specialists like herself. Recently, Lata has been trying to identify habitats in West Africa that might be threatened.

Although Lata agrees that more should be done to conserve biodiversity around the world, she is optimistic about the future. And she feels that careers like hers can go a long way toward protecting wild species. "This is a special profession," Lata says. "Enter it with conviction!"

A Friend of the Forest

In high school, Leslie Weldon spent a couple of summers in the Blue Ridge Mountains of Virginia working with the Youth Conservation Corps. She loved every minute of it—from helping to build and repair trails to cleaning up campgrounds along the scenic Blue Ridge Parkway. Learning about conservation and working outside in such a beautiful setting were, she says, the best parts of her experience. And she chose a career in conservation as a result.

These days, Leslie is working to conserve America's forests as a **biologist** at the U.S. Department of Agriculture's Forest Service headquarters in Washington, D.C. Before working at the headquarters, Leslie was a **forest ranger** in Montana's Bitterroot National Forest. A big part of her job there was to figure out what Americans wanted and expected from their national forests—things like clean water, wood products, abundant fish and wildlife, or places to hike and camp. Leslie thinks that her skills as a biologist are important as she helps the Forest Service manage the forests, but even more important is her ability to listen carefully to the public. As she says, "National forests belong to all of us."

Leslie Weldon

A Dream Come True

When Lynn Margulis was a child, she wanted to be an explorer. She dreamed of visiting unmapped tropical islands, jungle pyramids, and all kinds of other exciting sites around the world.

Today, as a **biologist**, Lynn gets to do a lot of exploring. She also gets to do something else she dreamed of doing as a child: writing. But she doesn't visit or write about uncharted islands or ancient monuments.

Lynn Margulis

Instead, she explores something equally fascinating and mysterious: the inner workings of the smallest living things—microscopic creatures. She studies the tiny creatures in their habitats: mud flats, salt marshes, puddles, and ponds. "Science for me is exploration," says Lynn, who adds that "No scientific work is complete if it has not been described and recorded in an article by the scientist herself."

Awake by 5:30 most mornings, Lynn bicycles to her research laboratory at the University of Massachusetts. In addition to conducting research and writing about it, Lynn is a **professor** at the university. And she also helps develop and write science teaching materials for students at levels ranging from elementary to graduate school.

Lynn feels that two different things have contributed to her success. One is her close relationship with her family, especially her four children. The other is reading. "I read nearly everything in sight," she says, "from poetry and novels to bottle labels, train schedules, and recipes!"

Say Cheeeese!

Did you know that Kodak's 35mm film cans contain 20 percent less plastic than they used to? Or that more than 150 million Kodak Fun Saver™ one-time-use cameras have been recycled since 1990? Maria Rasmussen knows! She's an **environmental engineer** and a manager at Kodak, and she works on helping Kodak protect both the environment and the health and safety of its employees, customers, and the communities where the company operates.

Working on environmental issues for companies can provide many opportunities and challenges. For example, throughout her career, Maria has helped Kodak's factories cut back on waste and properly manage the waste they do produce. She has looked at factories around the world to ensure that they meet both government regulations and Kodak's own set of environmental standards. And she has come up with ways of rating the environmental standards of Kodak suppliers.

Maria also works with other companies to share information and help the film industry as a whole do a better job for the environment. What is her advice for students interested in pursuing a career such as hers?

"Have fun, join clubs, and get involved!" she says. She also recommends "studying hard and taking as much math and science as you can, so you can understand the world around you and help protect it."

Maria Rasmussen

Making Waves

Suzanne Iudicello is not your typical lawyer. In fact, the majority of her clients don't even walk or talk. As an **environmental attorney**, Suzanne represents our nation's oceans and the diversity of life they contain.

Suzanne Iudicello

Because fish aren't exactly warm and fuzzy like pandas and wolf pups, Suzanne finds it tough to get people fired up about protecting them. As a result, she spends a lot of time and energy trying to boost the public image of sea life. She does this by speaking at special government meetings that deal with ocean-related issues, talking with representatives of the fishing industry, and working with conservation groups.

If you'd like to "make waves" as Suzanne does, she recommends getting a good basic education and plenty of experience. "Find interesting internships and volunteer opportunities that give you experience and that enrich and round out your academic training. And don't lose sight of what environmentalism is all about—protecting places you love. Get out in them and spend time on the water, in the marsh, in the mountains, in the woods."

Diving the Depths

Most people are terrified at the mere thought of being in the water with sharks. But Tundi Agardy hardly gives it a second thought! As a **marine biologist**, Tundi often finds herself diving into shark-infested waters.

Tundi first became interested in conservation while participating in a high school exchange program in the Caribbean. "One look at the Caribbean's tropical marine environment fixed my interest for a lifetime," she says.

Tundi is usually in the water at the first hint of daylight. Between dives, she writes notes on her observations and the data she has collected. She also teaches others about marine conservation and how to collect and analyze data.

Tundi feels that her biggest challenge is "to combine science and policy in a way that achieves long-lasting biodiversity conservation—conservation that benefits local people as well as nature." She insists that all kinds of experiences and education are important to working in conservation, and recommends that students "stay open-minded" and "try and find a way to do volunteer work so you can experience field research and conservation firsthand."

Tundi Agardy

Taking on Toxins

The first thing Theo Colborn remembers as a child is playing in a creek near the New Jersey farmhouse where she lived—turning over rocks, catching crayfish, and enjoying the sunshine. Her mother loved birds and flowers and gardening, and Theo learned to love them, also. "Being outdoors was a very important part of my life," she recalls.

Much later, Theo moved to a farm in Colorado with her husband and children. Life there was peaceful until the world's largest deposit of high-quality coal was discovered in the valley where the family lived. It was the energy crisis of the 1970s when oil was in short supply and mining boomed. Theo began to express her concern about how the mining was affecting Colorado's environment by speaking at open government meetings. She worked with local conservation groups until frustration at not being taken seriously motivated her—at the age of 51—to go back to graduate school.

Theo Colborn

Theo now works as an **environmental health scientist** at World Wildlife Fund in Washington, D.C. Her work has created widespread concern about the ways human-made chemicals may be affecting the environment. Her research is showing that certain chemicals can interfere with the development of wild animals before they're born if the chemicals get into mothers' bodies. It has also touched off new research into how these chemicals may be affecting our own health. Because chemicals have never been tested for these effects before, Theo's job now is to help the government come up with new tests to look for these effects. She says it's a tough job, but for the most part, "it's unbelievably stimulating."

Caring About People, Parks, and Wildlife

Kneel down, face the ground, and pray. That's exactly what Henri Nsanjama did when he was charged by a male silverback mountain gorilla on a recent visit to Rwanda. Henri, vice president of World Wildlife Fund's Africa and Madagascar program, has been working to protect Africa's biodiversity for more than 20 years. Born in a tiny village in Malawi, Henri had his first encounters with wildlife as a boy on hunting trips with his grandfather. But over the years, he began to notice that wildlife populations were dwindling while human populations kept growing and spreading. By the time he was 19, Henri was so concerned about what he saw that he knew he would make **wildlife conservation** his life's work. He went on to study wildlife biology, resource economics, and environmental management.

Henri has overseen projects designed to protect rhinos in Cameroon, elephants in Zambia, giant lobelia plants in Uganda, and lemurs in Madagascar. All these projects were designed to benefit local people. Henri believes that the key to conserving African wildlife is to find ways for local communities to "have a stake in it." One of the most challenging aspects of Henri's job is "making people outside Africa understand that

Africans often sacrifice their lives and property to conserve wildlife." During the past several years, Henri has helped train park wardens in Kenya to help curb poaching of rare rhinos.

By 4:45 AM, Henri is in the office before the vast majority of his coworkers to take advantage of the time to catch-up on paper work and the time difference in Africa. Although he admits that the pay isn't as high as it might be in some other fields, he says he receives "101 percent job satisfaction." And he adds, "I don't think I could have done anything better with my life than this."

Henri Nsanjama

Pigs, Pollution, and Politics

Every day, the governors of our 50 states and 4 territories have to make tough decisions about the environment. But luckily they have help. Robin Grove, a **natural resource policy expert,** helps these leaders make informed decisions by providing them with current information and best practices from around the country and the world.

Robin grew up in the beautiful state of Oregon—with sprawling desert plains, snow-capped mountains, fertile valleys, and scenic rocky coasts. In that rich environment, he got involved in politics and learned that elected leaders, like our governors, were faced with important decisions about how to protect the state's environment and meet the economic needs of the people. "I was also lucky enough to be in a state that led the nation in environmental thinking, with the first 'bottle bill,' people-friendly parks and green spaces, and creative urban planners. It taught me the value of thinking outside the box."

Today, Robin helps identify "best practices" in each of the states and territories. He and his staff then share these lessons with the governors so that every state can benefit. The issues he covers are varied—from how to clean up toxic waste sites to how to control pollution from big cattle, hog, and poultry farms. He also works on ways to produce more environmentally friendly electricity, to best manage state forests, and to ensure that poor communities are protected from pollution.

Robin believes that every person—from individual citizens to state governors—can help improve environmental quality. "Just get involved in what's happening in your community or state. And don't be afraid to ask questions of your political leaders." He also recommends that students learn the science behind the issues, become good writers and communicators, and keep up on the news. "Who knows," he says, "someday you might catch the political bug and run for governor of your state or territory!"

Robin Grove

Before the Interview

✓ When you call to set up an interview, introduce yourself on the phone. State your name, school, grade level, and the purpose of the interview.

✓ Set up an appointment far enough in advance to give you and the interviewee time to prepare.

✓ Carefully prepare your questions in advance. Limit the number of questions to about 10 or so. (Most people don't have time for long interviews, and too many questions will make it difficult for you to process all the information.)

✓ Find out if the interviewee would like a list of your questions in advance. If so, send them out as soon as you can.

✓ If you want to use a tape recorder, ask permission first. And make sure that it works and you know how to use it before your interview.

✓ If you are working in pairs, decide who will be asking the questions and who will be taking notes. (If the person taking notes thinks of additional questions during the interview, he or she can ask them. Try to make sure any new questions are brief and appropriate to the subject.)

During the Interview

✓ Be polite and considerate.

✓ Before you begin asking questions, explain how you will use the information.

✓ Ask your questions clearly and give the interviewee time to think and respond.

✓ Before you end the interview, thank the interviewee for taking the time to help you with your project.

✓ If you will be writing up the interview as an article, ask the interviewee if he or she would like a copy of it. If so, get the interviewee's address, and then be sure to follow through on getting the article to him or her as soon as you can.

✓ Ask for the interviewee's mailing address so that you can send a thank-you note or any other material after the interview.

After the Interview

✓ Send a thank-you note a few days after the interview.

✓ If you are working in pairs, meet with your interviewing partner soon after the interview to compare notes, impressions, and information.

DILEMMA 1: THE SAGA OF SAMMY THE GORILLA

Sammy is a western lowland gorilla that lives at the Davis Park Zoo in California. He's being considered for use in a breeding program as part of the Lowland Gorilla Species Survival Plan (SSP). SSPs are plans that are developed by teams of scientists, educators, veterinarians, and others. The plans work to maintain genetic diversity and to increase populations of threatened and endangered animals through captive breeding.

Each SSP team manages the breeding of a species to maintain a healthy and self-sustaining captive population. The SSPs also include research, public education, reintroduction, and field projects. Currently, there are 84 SSPs covering 136 species. One of those SSPs focuses on lowland gorillas, like Sammy.

Sammy currently lives with a female gorilla named Tiny. The two gorillas get along well, but they haven't produced any offspring because Tiny is unable to have babies. Because of that, the SSP management committee has recommended moving Sammy to the McCall Zoo in Ohio. There are three female gorillas at this zoo, and officials hope Sammy will be able to father offspring with one or more of them.

But plans to move Sammy have raised concerns. Sammy is the most popular attraction at the Davis Park Zoo, and many of his "fans" are upset that he might be leaving. Some people also feel that separating Sammy and Tiny is wrong. Hundreds of people have signed a petition stating that they will boycott the zoo if Sammy is moved.

The final decision about whether to move Sammy rests with a committee that consists of the director of the zoo and the staff at the primate house. Although zoos usually follow SSP recommendations, all of the controversy surrounding this move has forced the committee to reconsider Sammy's move. If your group were this committee, what would you do? To help you make your decision, we've provided the following documents:

- a letter from the Species Survival Plan Coordinator to the Director of the Davis Park Zoo

- a press release announcing the possibility of moving Sammy to the McCall Zoo

- a cover sheet that accompanied the petition signed by people against the move

Letter

James Donovan, Director
Davis Park Zoo
Cherryhill, CA 90081

Dear Mr. Donovan,

As you know, it is my job to coordinate the Species Survival Plan (SSP) for western lowland gorillas. You have the opportunity to help these endangered animals by being part of a breeding program designed to increase their population. Your participation would involve moving your male gorilla, Sammy, to Ohio's McCall Zoo.

After conducting extensive research and computer modeling, the SSP management team believes that Sammy's genes would greatly enhance the captive breeding of the western lowland gorilla species. The McCall Zoo has three females that are capable of bearing young. Because Sammy is one of only a few captive gorillas that were born in the wild, it is likely that he is not closely related to these captive-bred females. Having a different genetic line makes him very valuable to this breeding program because it lowers his offsprings' risk of genetic problems caused by inbreeding.

I understand that your zoo does not have space for another gorilla, so it would not be practical to move any of the females from Ohio to Davis Park. I also think it would be disruptive to Ohio's females if your female gorilla, Tiny, accompanied Sammy. In addition, I hope you agree that it would be in the best interest of all parties if Sammy's move was permanent. By fathering offspring for many years to come, Sammy stands to contribute greatly to efforts to save his species from extinction. And because moving can be extremely upsetting and traumatic for gorillas, staying at his new home would also seem the most humane course of action.

As for Tiny, perhaps a male from another zoo could be moved to Davis Park to live with her. Since gorillas are very social animals, it would be better for Tiny to have a partner.

I hope we will be able to make arrangements to move Sammy soon, and I look forward to speaking with you about this matter.

Sincerely,

Tom Buford

Tom Buford
Species Survival Plan Coordinator
Western Lowland Gorilla

Davis Park Zoo

Press Release

Sammy to be moved to Ohio's McCall Zoo

Sammy, a male western lowland gorilla and a popular resident of the Davis Park Zoo, may soon be moved to Ohio's McCall Zoo. The move would enable Sammy to take part in a special program called a Species Survival Plan (SSP).

Species Survival Plans are designed to help conserve endangered species. So far, SSPs have been developed for more than 80 endangered species living in zoos in the United States. One of the goals of many SSPs is to breed species such as western lowland gorillas in captivity. Doing that increases both the numbers of animals and the genetic diversity of zoo populations. Other goals of SSPs may include protecting habitat in the wild, maintaining genetic diversity, and educating the public about endangered species.

Sammy is currently housed with Tiny, a female gorilla who is unable to produce offspring. If Sammy is moved to Ohio, he will be slowly introduced to the McCall Zoo's three female gorillas. Zoo officials hope Sammy will father offspring with one or more of these females.

Sammy's contribution to the captive population is particularly important because, unlike most gorillas in zoos in the United States, he was born in the wild. Consequently, his genetic makeup is different from that of most other zoo gorillas in this country. As a result, Sammy will be able to help strengthen the gene pool of gorillas living in zoos.

In the wild, western lowland gorillas live in several African nations, including Cameroon, Congo, and Nigeria. They usually form social groups made up of one dominant male, several females of breeding age, and young gorillas of different ages. Since gorillas do not mate for life, males typically father offspring with a number of different females.

Lowland gorillas like Sammy are endangered because the forested habitat where they live is being logged and cleared to make way for farms. Western lowland gorillas are also hunted illegally.

Petition

Society for the Humane Treatment of Animals

We the undersigned are opposed to any plans that would result in moving Sammy, the Davis Park Zoo's male gorilla, to the McCall Zoological Park in Ohio. In light of the close relationship that has developed between Sammy and Davis Park's female gorilla, Tiny, we believe that separating the two animals would be an act of cruelty. The bond between the two gorillas is obvious. They often cuddle and sleep in each other's arms and they are always together. Neither Sammy and Tiny's close relationship nor the stress and trauma to Sammy of being moved across the country is being considered in the decision-making process. A keeper who works closely with Sammy has expressed concern that Sammy has just recently begun to adjust to his current situation and has become quite attached to Tiny. In light of these conditions, we believe that such a move could endanger his health or even his life.

In addition, we are not convinced that moving Sammy will benefit his species, as has been stated by those in favor of the plan. He is described as a shy animal, and there is no guarantee that he will establish bonds with the three females at the McCall Zoo. Even if he does adjust to the females and eventually produce offspring with them, we believe that the claims of his possible contribution to gorilla survival are exaggerated. Since it's so expensive to actually release captive-bred gorillas into the wild, we believe that Sammy's immediate and future offspring will probably remain in zoos.

In our opinion, money is the driving force behind the push to move Sammy. It's no secret that baby gorillas can greatly increase the numbers of visitors to a zoo. Having more visitors means not only more money collected at the entrance gate, but also more money spent on souvenirs, food, and other zoo merchandise. The Davis Park Zoo is clearly under financial strain, and they have already stated that they do not have any room to bring in Tiny. They also don't have money to create more space. We believe it is immoral and offensive to be more concerned with money than with the quality of Sammy's life.

For these reasons, we cannot support plans to move Sammy. Please be advised that moving Sammy will result in a permanent boycott of the Davis Park Zoo by those who have signed this petition.

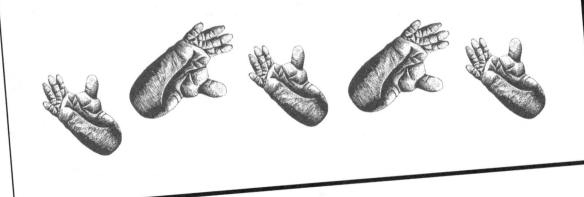

DILEMMA 2: SAVING AN ENDANGERED PARROT

The orange-necked parrot* is an endangered species found on only one island in the Caribbean. This island provides the unique blend of plants, climate, elevation, and food that orange-necked parrots need to survive. Just one flock of the birds, made up of 75 individuals, remains.

Thousands of orange-necked parrots once lived on the island, but the birds have been steadily dying off over the past two decades. Scientists are not certain what's causing the birds' decline, but they suspect it's due to a combination of habitat loss, drought, and pesticide use on local farms.

Some scientists want to try to capture the remaining parrots and breed them in zoos. When the problems leading to the decline have been addressed, the scientists would introduce some of the captive-bred birds back into the wild. It's not certain if orange-necked parrots will produce young in captivity. But other species, such as the thick-billed parrot, have been successfully bred in zoos.

Other scientists want to leave orange-necked parrots in their natural habitat and find ways to protect them there. They feel that removing the parrots from the wild might mean that the birds' habitat would no longer be legally protected. (The island has laws that prohibit an area from being developed when an endangered species lives there.)

What course of action do you think scientists should take to protect orange-necked parrots? Why? We've provided data about the thick-billed parrot captive breeding program. We've also included some other things to think about as you make your decision.

* The orange-necked parrot is a fictional species.

Fish and Wildlife Agency
—Memorandum—

Date: July 1995

From: Chief, Office of Endangered Species

To: Regional Director

Subject: Update on Thick-Billed Parrot Program

The captive breeding program for thick-billed parrots is going well, as you can see from the data below. These parrots are quite easy to breed in captivity, and one of the program goals has been to produce enough birds for a healthy population. Some of the birds have become a part of the reintroduction program in Arizona, one of the areas where thick-billed parrots originally lived.

However, the reintroduction program has not been as successful as the captive breeding effort. There have been problems with drought, predators, and shortages of pine cones, which are the birds' favorite food. For example, in November 1991, 18 captive-bred thick-billed parrots were released in southeastern Arizona. Within a short time, half the birds had died. Some had been eaten by hawks; others starved or died from disease. By January 1992, another four birds had died from predation.

Staff scientists think that larger flocks of parrots might be better able to defend themselves against predators. Therefore, future releases will involve larger groups of birds. The releases will also occur at a time of year when there is a lot of food. In addition, scientists are studying how to better prepare captive-born parrots for life in the wild. Many scientists think that some of the released parrots died because they didn't know how to find food and protect themselves in the wild.

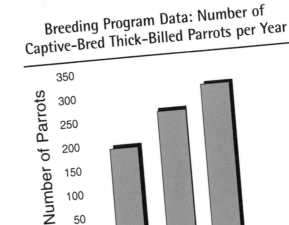

Breeding Program Data: Number of
Captive-Bred Thick-Billed Parrots per Year

Things to Think About

✓ Because orange-necked parrots have never been kept in captivity, there are many uncertainties about how to best care for them in a zoo. For example, scientists don't know much about the birds' nutritional and space needs. Generally, parrots breed well in captivity. However, there's no guarantee that orange-necked parrots would do well. And even if the parrots did breed, scientists don't know if the young would be able to adjust to living in the wild if they were reintroduced.

✓ The factors that caused the parrot population to shrink might still be there when the young are released. Scientists aren't sure what these factors are, but they suspect a combination of pesticide use, habitat loss, and drought is to blame. Scientists need to do more research to better understand these and other possible causes of the birds' decline—but research takes time and money.

✓ If the parrots are left in the wild, the decline would probably continue and could easily lead to extinction of the orange-necked parrot.

✓ If pesticide use, habitat loss, and drought are the reasons orange-necked parrots have died off, then the birds might make a comeback if pesticide use and habitat loss decrease, and if the drought ends.

✓ Some people have suggested that only some of the remaining orange-necked parrots be removed from the wild to start a captive flock. But many scientists are reluctant to try this. They point out that, like thick-billed parrots, orange-necks in the wild might be safer in large flocks. Removing some of the wild birds might leave those still living in the wild more vulnerable to predators.

✓ The California condor captive breeding program has been successful, although it was controversial at first. In the late 1980s, all of the remaining condors—a total of 27 individuals—were taken out of the wild to breed in captivity. Despite setbacks, many captive-bred condors have been successfully returned to the wild. As of October 1998, there were 104 condors in captivity and 46 in the wild. However, scientists point out that there are many differences between the condor situation and that of the orange-necked parrot—and between the two species themselves.

✓ Deciding to leave the parrots in the wild requires a program focusing on community education and reserve management. Large sums of money would be needed to set up an official preserve, to develop a recovery plan/project, and to create public education programs.

How Can We Protect Biodiversity?

As the director of research at a zoo in Kentucky, it's your job to decide how this year's research money ($300,000) should be spent. You have asked two of your best researchers to give you their ideas on how to spend money over the next three years to help the endangered Brooks antelope.* Your zoo has several of the animals.

The Brooks antelope is endangered for two reasons. One is illegal hunting. The other is that the antelope's forest habitat is steadily being cleared by people who need the land to grow their food. Scientists have tried for years to breed the animals in captivity, but so far not one of the attempts has been successful.

On the basis of the following two written statements—one from each of your researchers—how would you spend the money? Why? What other information would you need? What questions would you ask each researcher?

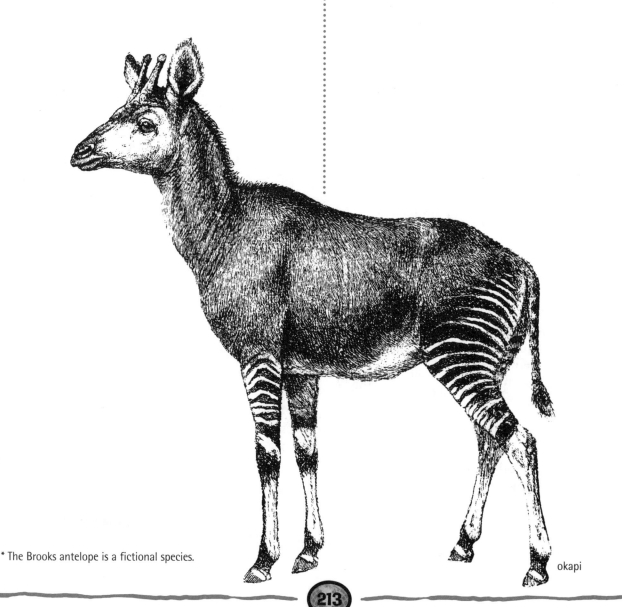

* The Brooks antelope is a fictional species.

okapi

Statement 1

As you know, female Brooks antelopes have not had successful pregnancies in captivity. Consequently, I feel that the species is a perfect candidate for an embryo transplant procedure. If the transplant is successful, it would result in the first Brooks antelopes born in captivity. Below is a description of how the procedure would work.

Using eggs from a female and sperm from a male Brooks antelope, we would create embryos in the laboratory. Embryos would be frozen until ready for use. Then, at the appropriate time, we would implant the embryos into rusty antelope females, which are closely related. If we achieve successful implantations, the rusty antelopes would give birth to Brooks antelopes after a pregnancy of five months.

Rusty antelopes seem to be the logical choice for this procedure for several reasons. First, our zoo has many rusty females. Second, unlike Brooks antelopes, rusty antelopes usually have successful pregnancies in captivity, which produce healthy offspring. And third, the rusty antelope species is not endangered. So, if an implantation doesn't work and causes a female rusty to get sick or die, the loss wouldn't be as crucial to the species as it would be if there were only a few of the animals left.

As you know, this procedure has worked well with other antelope species. For example, eland antelopes have successfully given birth to bongo antelopes, another endangered species that is difficult to breed in captivity. Of course, the exact details of the procedure will have to be worked out in advance.

I estimate that the entire breeding project, including the research needed to make it a success, will cost about $300,000. There are several advantages to the project, including the following:

- The project would generate publicity that could help educate the public about Brooks antelopes—and perhaps about endangered species in general.

- If the project is successful, the success may make it easier for us to raise the funds necessary to conduct more research on Brooks antelopes and their needs.

- We may be able to apply some of the discoveries we make during the procedure to the captive breeding of other species.

- If Brooks antelopes become extinct in the wild, our captive breeding project would help ensure that they would still exist in captivity. We might one day be able to release the captive-bred animals back into the wild.

Statement 2

I believe we need to protect Brooks antelopes in the wild. And to do that, we need to protect their habitat and prevent illegal hunting. Therefore, I propose that we launch a project that would focus on two very important tasks: research and education. The research component of the plan would involve an ongoing study of wild Brooks antelopes. The education portion would be aimed at reducing illegal hunting and habitat loss by helping local people learn about the antelope and its plight. Below is a brief summary of each of these areas.

Research

As you know, new farms are dividing the antelope's habitat into small, disconnected patches. Currently, we don't know much about the behavior and needs of Brooks antelopes in the wild, and so we have trouble recommending the best ways to help protect them and their habitat. Therefore, I recommend that we first study the animals in their natural habitat. Later, we can use what we learn to determine the best course of action to take. For example, we might discover that a special reserve free from farming activities should be set aside for the animals. Or we could find that changes in local farming practices—such as rotating fields every 10 to 15 years so that former fields have a chance to grow back into forest—would be what's necessary to protect the antelopes.

Education

For hundreds of years, tribal people depended on Brooks antelopes for food and other products. Although many people living in the animals' habitat now rely on cattle for their food, some still prefer to hunt the antelopes. A new law has made it illegal to kill the animals—but many local people resent this law because they feel that it doesn't respect their customs and traditional hunting practices. As a result, the law is often ignored, and the animals continue to be hunted.

Helping the local community start a public education program about the Brooks antelope would be a major component of our program. The people currently have little knowledge about how their hunting practices affect the antelopes' population or about the antelopes' habitat as a whole. In addition, scientists as well as local people have limited understanding of how their farming methods affect the animals' habitat. If an understanding of and concern for the antelope is developed through an education program, the local people may be willing to change their practices to save the animal from extinction—and also to provide a long-term source of sustainably harvested antelope meat for themselves.

Statement 2 (Cont'd.)

Budget Needs and Benefits

The entire research and education program would cost about $300,000. Following is a list of some of the program's benefits:

- Learning more about the antelopes' habitat needs will help protect the habitat itself. The more we're able to protect the antelopes' habitat, the better the chance that the habitat will still exist if and when the time comes to reintroduce captive-bred antelopes into the wild.

- An education program would give the local people a better understanding of how their farming and hunting practices affect the Brooks antelope. It would help protect not only the antelopes themselves, but also the habitat as a whole.

- Information gained through studying Brooks antelopes in the wild may also help in learning how to successfully breed them in captivity.

- The research component of the program could help in learning about other species that live in the same habitat as Brooks antelopes. As a result, we may discover the best way to protect not only the antelopes, but certain other species as well.

- Projects that help protect animals in the wild are good publicity for the zoo and can help with fund raising. They show that the zoo is committed to conservation in the field. These projects also provide the public with specific activities they can donate money to.

- Field programs are more cost effective than high tech, artificial breeding methods and processes.

DILEMMA 4: PLANTS AND PROFITS

A s the Wilson Botanical Garden's chief botanist, you are often part of a group that travels to the tropics to collect plants and their seeds. The botanical garden uses some of these plants and seeds to grow new specimens for public display. Once the plants are brought back to the botanical garden, researchers also study many of the collected plants.

Recently a researcher discovered that one of the plants collected by your team contains a compound that may be useful in treating cancer patients. If a medicine is developed from the plant, it could save many lives. A well-known drug company has already expressed an interest in trying to use the newly found chemicals to make a new medicine—but it takes a lot of time and money to do so.

Soon after the discovery of the plant's medicinal value was announced, you received a letter from an official in the country where your team collected the plant. In her letter, the official stated that her government expects to be given a share of the money from sales of the medicine once it has been developed.

You understand this point of view, but you also realize that researching and developing new drugs can be very expensive and risky to drug companies. If a company were to share a large part of the money from sales of a new drug, it would take the company longer to recover the money that it had spent on researching and developing the drug. You're concerned that the drug company that's interested in "your" plant might lose interest if it would have to share a large portion of the money from the medicine it develops. And in turn, that action would mean that a life-saving medicine might never be developed.

Using the information in the letter from the government official and a letter from the drug company, come up with some recommendations that could help solve the dilemma. Also, what questions would you want to ask indigenous people who live in the area where the plant grows naturally? What questions would you ask the drug company officials? What information is missing?

Letter from the government official

Leader, Bioprospecting Team
Wilson Botanical Garden

To Whom It May Concern:

We recently received a letter from your botanical garden stating that a plant collected in our country contains chemicals that may be useful in treating cancer patients. While we are thrilled to hear that one of our native plants could prove helpful in the struggle to combat this life-threatening illness, we have some concerns.

For decades, developed nations—particularly the United States—have sent researchers to collect plant specimens in our country. Usually the plants are not economically valuable. However, a drug company in your country now stands to make a large profit when one particular plant is made into a medicine. Our government feels that, because the plant originated in our country, we should be given a significant share of the profits.

As you may know, our country is not a wealthy one. We have little money to pay for forest conservation, yet affluent countries like yours have been pressuring us to conserve our tropical forests and endangered species. One of the best ways we can conserve our forests is to show our citizens that the forests can be more valuable if left standing than when chopped down for lumber or to make way for agriculture.

If our citizens can be shown that there are plants in the forests that this nation can profit from, our government will have an easier time passing laws that protect the forests. Currently, our people have no choice but to try to make a living by cutting the forest to grow food and to sell the trees. However, if we were able to receive income from the plant sales, the money could be used to develop forest-friendly businesses, non-timber industries, and training programs. Another idea would be for the company to start a technology transfer program where scientists would share information with our scientists, doctors, and farmers. All of these actions would encourage our people to help protect and maintain the forests.

We cannot continue to give away our valuable plant resources without substantial compensation. If we do not receive a large share of the profits from medicines developed from our plants, we will have no choice but to prevent U.S. researchers from collecting plants and seeds in our country.

I am also sending a copy of this letter to the drug company that has shown an interest in our medicinal plant. I hope that you both will take our views into account.

Sincerely,

Maria Lugo

Maria Lugo
Department of Natural Resources

Letter from the drug company

Maria Lugo
Department of Natural Resources

Dear Ms. Lugo:

We received your letter about the medicinal plant that researchers from the Wilson Botanical Garden collected in your country. We have agreed to share a small portion of the gross profits with you and plan to do so, should the medicine prove to be effective and profitable. But the possibility of sharing a larger percentage of the profit may make it impossible for us to develop the drug at all.

At this time, our research shows that a chemical in the plant seems to help shrink cancer tumors in laboratory rats. It will take years of expensive research to find out whether the effects would be similar in humans, what the proper dose would be, and what the side effects may be. (The chance of developing a viable drug from a discovery such as this is about 1 in 5,000.) It will also take time and money to get approval from our government to sell the medicine once we develop it from the plant. Finally, even if permission is obtained, it is impossible to predict how well it will be received by the medical community and whether we will even recover our costs, let alone make a profit.

You must understand that it is very costly to develop a new drug. In fact, the average cost of developing a successful drug in the United States is $231 million. Often we don't earn enough money to cover our research and development costs until years after we start selling a drug. To stay in the business of producing life-saving medicines, we need to reinvest the money from sales. The income will help us recoup our costs and fund this and future endeavors to develop life-saving medicines.

Sincerely yours,

Mitchell Brown
Research Director
ABC Drug Company

You are the curator of a new aquarium in a large city. Your exhibit development team (consisting of scientists, educators, exhibit designers, and public relations specialists) has just come to you with a dispute about whether a coral reef exhibit should be developed. Two of the four members are opposed to the idea and two are in favor of it. Using the following information, how would you resolve this problem?

The Case Against a Coral Reef Exhibit

Exhibit team members Malcolm and Wendy feel that having a coral reef exhibit is a bad idea because such an exhibit would include many species of tropical marine (saltwater) fish and other coral reef species. They're concerned that the display of colorful marine creatures could encourage visitors to set up saltwater aquariums in their homes. While having an aquarium full of beautiful marine fish and other species may seem like a nice hobby, Malcolm and Wendy point out that millions of reef animals are taken from the wild and sold in the pet trade each year.

Currently, there is not enough information on whether the collection of marine reef species for the pet trade is causing populations of these species to decline overall. But some of the methods used in collecting are damaging to coral reefs and their species. For example, chemicals such as cyanide are illegally used in certain countries to stun and capture reef fish. These drugs and chemicals often damage the coral itself, as well as other reef species. In addition, fish captured with the use of chemicals usually die in a matter of weeks or months.

Malcolm and Wendy believe that, instead of a live display, the museum could develop a colorful, creative, and engaging exhibit to inform visitors of the problems facing coral reefs. It could highlight the reasons the aquarium decided against using live specimens in this exhibit, and it could have a great impact on the public.

The Case for a Coral Reef Exhibit

Exhibit team members Mochita and Gary think that the educational value of including a coral reef exhibit far outweighs the risk of encouraging people to buy marine species for home saltwater aquariums. They point out that the dazzling colors of the reef species will draw a lot of people to the exhibit, and they believe that many of these people will read the accompanying signs that will be part of the display.

Besides generally educating people about coral reefs and their species, Mochita and Gary have suggested that the signs include information about the trade in wild tropical fish and its negative impact on coral reef ecosystems. Video footage could be used to compare and contrast a pristine, healthy reef with a damaged one. The video could also include a section showing the collecting and other practices that led to the damage. Providing a flow chart of how these illegally captured species move into the U.S. market would help to better educate the public about this issue.

Mochita and Gary believe that, through a live exhibit, people will be more likely to care about coral reefs and species that live there. They stress that a live exhibit is a good way to display a whole ecosystem and to interpret the wonders of the ocean in a more meaningful way than with a nonliving exhibit. And they feel that in the long run this could mean that coral reefs and tropical marine fish will be more likely to be protected.

DILEMMA 6: DIFFERENT POINTS OF VIEW

A month ago the Dixon City Natural History Museum opened a large exhibit on biodiversity. In general, the exhibit has been warmly received by Dixon City residents and tourists. However, there have been a number of angry letters concerning some of the language and ideas presented in the exhibit. It is still unclear just how many visitors share the views expressed in the letters, but as director of the museum you are worried that the controversy could grow. It is your job to decide how to deal with the concerns of the letter writers. Should you change or add information to the exhibit signs? Should you wait to see if more letters come in? Are there other ways to handle the situation?

Use the examples of text from the biodiversity exhibit and the following letters (published in the *Dixon Daily News*) to decide how to deal with this dilemma.

Exhibit Text Samples

- Biodiversity is more than just a huge collection of birds, bugs, slugs, cats, bats, cacti, and the millions of other organisms that inhabit our planet. It's also a crucial part of Earth's ability to sustain life. Species and the ecosystems they inhabit help to maintain our atmosphere, to keep our soil fertile, to purify our water, and to generally keep things running smoothly. By allowing biodiversity to decline, we're threatening the very systems and functions that allow life to exist on our planet. We must, therefore, do everything we can to protect biodiversity.

- Across the globe, people are using natural resources faster than natural systems can replenish them. This is especially true in developed nations such as the United States, where we consume tremendous amounts of oil, trees, water, and other natural resources each day. Most experts agree that, if we are to protect biodiversity, we must reduce our consumption of these resources and help protect habitats worldwide.

- People are working to protect biodiversity in a variety of ways. One result of these efforts is the development of legislation designed to protect species and ecosystems. The Endangered Species Act is one example of such legislation. Since its enactment in 1973, this law has helped to protect species such as the bald eagle, the American alligator, the whooping crane, and many others. But protecting species—and biodiversity in general—will take more than laws. It will also take public awareness—and, most important, a willingness on the part of citizens, communities, businesses, and governments everywhere to change old patterns of resource use. Everyone has a role to play in protecting our planet's amazing diversity of life.

Letter to the Editor 1

Dear Editor:

My family and I recently visited the Dixon City Natural History Museum's new exhibit on biodiversity, and we were disappointed by some of the biased viewpoints. For example, the text accompanying the exhibit includes the following statements:

- "We **must** . . . do everything we can to protect biodiversity."
- ". . . we **must** reduce our consumption of natural resources."
- "Everyone has a role to play in protecting our planet's amazing diversity of life."

When I visit a museum exhibit, I do not feel that I should be told what I **must** do. Nor do I want my children to be told what they must do. Also, people should realize that when our country produces and consumes lumber, minerals, and agricultural products, jobs are created and the economy grows. I work for a paper company and am one of those people who has a job thanks to "the consumption of natural resources." It appears that whoever created the exhibit had no idea how some of their suggestions for protecting biodiversity will affect the economy and the livelihood of families like mine.

Another complaint I have about the exhibit is the way it portrays the Endangered Species Act. While it's true that the Act has saved species such as the bald eagle from extinction, nowhere does the exhibit mention the effects this legislation has had on people. My job has been threatened a number of times because of the Endangered Species Act. It took my company years to get a permit to continue with our work because the land we wanted to use was also the habitat of an endangered plant. And it was on private—not public—land! The exhibit leaves out these consequences of the Endangered Species Act.

I have nothing against nature, but I do think the exhibit is too preachy and ignores the needs of many people. And I can tell you I am not the only one with this viewpoint.

Roger Drake

Letter to the Editor 2

Dear Editor:

I would like to take this opportunity to respond to Roger Drake's letter that appeared in the newspaper last week regarding the new exhibit at the Dixon City Natural History Museum. Like Mr. Drake, I have also visited the exhibit. But unlike him, I feel that the exhibit was not strong enough in its message of conservation. Many plants and animals are becoming extinct every day, we continue to pollute our air and water, and we are using our natural resources at an ever-increasing rate. If all of us do not do everything we can to protect biodiversity, we will be in big trouble in the future. I firmly believe that everything in nature is connected to everything else.

I would also like to say something about the Endangered Species Act. Does Mr. Drake think he has no personal responsibility for protecting endangered species? If no one is willing to take responsibility, then we could continue to lose endangered species in the future. Personally, I believe that the act is not strong enough because it focuses on saving individual species instead of protecting whole ecosystems.

We need to do more, not less. Please don't let Mr. Drake's views prevent you from your important role of educating the public about biodiversity.

Cecilia Smith

Cecilia Smith

Everyone has a role to play in protecting our planet's amazing diversity of life.

unning out of water may seem hard to imagine. After all, water comes out of your faucet every time you turn it on. But some parts of the United States are increasingly facing water shortages and other water issues. And even for a city that's using water wisely, everything can change when drought strikes. What would you do to ensure that your town has enough water now and in the future, regardless of the circumstances?

One California coastal city, Santa Barbara, had to deal with a water emergency when it was hit by one of the most severe droughts ever recorded. After learning about the situation that confronted Santa Barbara, your team is going to work together to develop a plan that might have helped Santa Barbara survive its water shortage. At the end, you'll be able to compare your plan with what Santa Barbara actually did until the rains finally came.

Where does Santa Barbara's water come from?

Almost all the water for the approximately 90,000 people in Santa Barbara comes from the Santa Ynez [EE-nez] River. The city owns its own reservoir, called the Gibraltar Reservoir, where rainwater and water from the Santa Ynez River are stored. Down river from the Gibraltar Reservoir is Lake Cachuma. Lake Cachuma's water is shared by Santa Barbara and other nearby cities and towns. Santa Barbara also gets some of its water from under ground, but the contribution made by this water (called groundwater) is small. Before the drought hit, there was talk of expanding the Gibraltar Reservoir to store more water for the city of Santa Barbara, but the plan had to be shelved because it would have flooded habitat for an endangered bird called the least Bell's vireo.

The Drought

The drought that caused water shortages in Santa Barbara began in 1986. By 1989, the Gibraltar Reservoir was empty. If the drought continued, experts predicted that the volume of water in Lake Cachuma would be down 45 percent by 1990 and that the lake would be empty by 1992. Clearly, water officials needed to do something, but they were hesitant to take drastic measures right away. If they spent a lot of money to increase the dam size at Lake Cachuma or to pipe in water from elsewhere, the projects would not even be complete before the drought ended. Also, they were worried that calling for tough water conservation before it was absolutely necessary might anger residents, or that a false alarm would make residents less likely to trust water officials' decisions in the future.

So the officials in Santa Barbara decided on a plan with three stages. Each stage had a different set of projects or restrictions. As the drought got worse, officials planned to move from Stage I through Stage III.

You will be reading about some of the things that water officials considered when they were deciding how to save water or how to bring more water to the residents of Santa Barbara. Using these ideas, see if your team can develop a three-stage plan of your own to get Santa Barbara through the drought. Make sure you are able to explain your reasons for choosing certain options and how you might get around some of the problems described. You might also want to think about how water could continue to be used wisely once the drought was over.

There are a number of ways to reduce water use. To achieve those reductions, you can allow people to act voluntarily or you can impose regulations that people must follow in order to avoid some type of punishment, such as a fine. If you choose to use water regulations, your plan should include ideas about how to enforce them. Here are some things to focus on when considering how to reduce water use:

Watering Plants: For people with lawns and gardens, you can set limits on when and how they can water their plants. For example, watering early in the morning or in the evening means less water will be lost to evaporation. And less water is lost by using hoses or pipes that are allowed to drip slowly into areas being watered (called drip irrigation) than by using sprinklers. Keep in mind that if plants don't get enough water they may not survive. And, in general, it would cost more to replace dead trees and shrubs than to replace a lawn.

Filling Swimming Pools: Many people in Santa Barbara have swimming pools. One way to reduce water use would be to keep people from adding water to their pools. Also, putting a cover over the pool when it is not being used would keep water from evaporating.

Washing Cars: Washing cars uses a lot of water, particularly if the water is left on when the car is being soaped up. Some commercial car washes are designed to reuse the same water several times.

Using Water to "Sweep": Using a broom rather than a hose to remove dirt and leaves from paved areas helps to conserve water. When water is used, much of it runs into the street and is wasted.

Using Showers, Faucets, and Toilets: Taking long showers or leaving a faucet running (when brushing your teeth, for example) wastes water, as does unnecessary flushing of toilets or using toilets that require large volumes of water. People can buy and install "low-flow" shower heads and toilets that use less water. If your team wants to encourage people to replace their shower heads and toilets, how can you get people to buy them? People can also place a rock, a brick, or a bag of water in the toilet water tank to reduce the amount of water it takes to fill the tank. They can also forgo showers that aren't necessary.

Offering Water in Restaurants: Many restaurants place a glass of water in front of their customers whether or not they ask for it. If they don't drink the water, it is thrown away. At some restaurants, you get water only when you ask for it.

Running Fountains: Many people would agree that fountains are beautiful but not necessary. Fountains can be turned off completely or the water can be used over and over again. Even when fountains recycle water, their use wastes water through evaporation.

Restricting Hotel and Motel Water Use: Out-of-town visitors might not know about the drought. Should hotel and motel guests have their water use restricted? How will guests know what to do?

Raising Water Rates: One way to reduce water use is to raise the cost of using water. However, some people won't lower their water use because they aren't worried about how much it costs, they have important reasons for using the water, or they are already conserving as much water as they possibly can. There are a few things your team should think about if you are considering raising water rates. How will you deal with households that can barely afford to pay current water rates? How will you deal with households that have a lot of people? For instance, a doubling of rates for a household with seven people would be a lot more money than for a household with only two people.

Reclaiming Water: Waste water can be partly cleaned and then used to water large lawns and planted areas like those you would find at golf courses, schools, and parks. This "reclaimed" water doesn't smell bad or cause health problems for people using the areas sprayed with it. Water reclamation reduces demand on drinking-water supplies.

Desalinating Water: Desalination takes salt from water and makes it into freshwater. Usually this process is done at a desalination plant by removing the salt from the water through distillation, filtration, or reverse osmosis. Desalination may seem like an ideal way for a coastal town like Santa Barbara to get a new supply of water, but there are some drawbacks. While building a desalination plant costs about the same as some other major water projects (like building new dams or piping in water from elsewhere), it is more expensive to produce water once one is built. Also, in most cases it takes a minimum of 18 months to build a plant, so this option is not something that can relieve a drought immediately.

Providing Public Education and Information:
Whatever plan your team develops, the residents will need to know what's happening. How will you let people know about new water use restrictions, desalination, reclaimed water, or anything else you might have included in your plan? Make sure to include your ideas for educating and getting information to Santa Barbara's residents.

Notes:

Notes:

Notes:

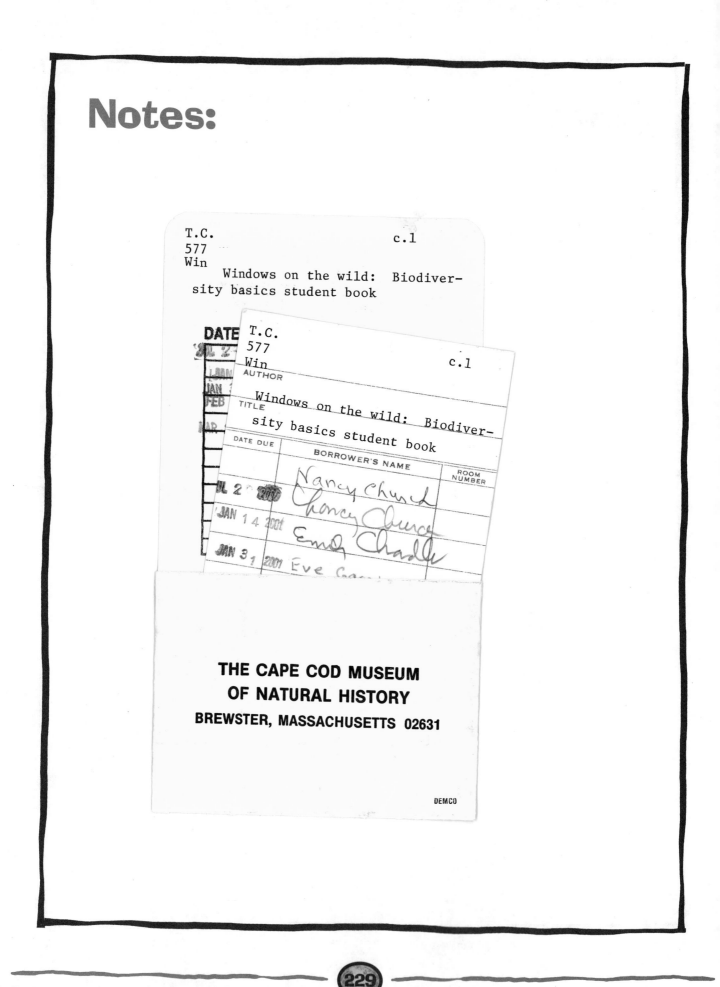

T.C.   c.1
577
Win
        Windows on the wild: Biodiver-
sity basics student book

DATE

T.C.
577
Win   c.1
AUTHOR

        Windows on the wild: Biodiver-
sity basics student book
TITLE

DATE DUE	BORROWER'S NAME	ROOM NUMBER
IL 2 7 2000	Nancy Church	
JAN 1 4 2001	Nancy Church	
JAN 3 1 2001	Emily Charly	
	Eve Gra	